The Last Great Miller

The Four Wheel Drive Indy Car

Other SAE books on this topic

The Golden Age of the American Racing Car
By Griffith Borgeson
(Order No. R-196)

The Last Great Miller

The Four Wheel Drive Indy Car

Griffith Borgeson

SAE INTERNATIONAL

Society of Automotive Engineers, Inc.
Warrendale, Pa.

Library of Congress Cataloging-in-Publication Data

Borgeson, Griffith.
The last great Miller : the four wheel drive Indy car / Griffith Borgeson
p. cm.
Includes bibliographical references and index.
ISBN 0-7680-0500-0
1. Miller, Harry A. (Harry Armenius), 1875-1943--Correspondence. 2. Automobile mechanics--United States--Correspondence. 3. Indy cars--History--Sources. 4. Antique and classic cars--History--Sources. 5. Four-wheel drive vehicles--Design and construction--History--Sources. I. Title.

TL140.M54 .A4 2000
629.228'09--dc21 00-036542

400 Commonwealth Drive
Warrendale, PA 15096-0001 U.S.A.
Phone: (724)776-4841
Fax: (724)776-5760
E-mail: publications@sae.org
http://www.sae.org

ISBN 0-7680-0500-0

SAE Order No. R-244

Contents

Preface

This is Griffith Borgeson's last book. Griff died in 1997 shortly after completing the first draft. His text has been left unchanged, although a few explanatory footnotes have been added where there were indications that research was in progress. Some parts of the book do not flow as well as they would have had Griff lived to finish the book himself.

This is the story of the Four Wheel Drive (FWD) Special, a car conceived by Harry Miller, arguably America's most brilliant race car builder. Harry Miller convinced the Four Wheel Drive Auto Company to fund the project. Beginning with the initial letter from Harry Miller, through the completion of its racing years, the Four Wheel Drive Auto Company kept an amazingly detailed file on the car. The richness of this file containing a vast amount of personal correspondence allows this story to be told in detail and with unique insight. The extensive reproduction of these letters in the text of the book gives the reader a first-hand feel for Harry Miller, Walter Owen, and the other characters involved.

The FWD car ran at the Indianapolis Motor Speedway in the Indy 500 each year from 1932 through 1937. Its best finish was fourth in 1936, driven by Mauri Rose. The car resided in the FWD museum until 1948 when Bill Milliken, one of the best-known authorities on race car handling, and coauthor or the best selling work on the subject, *Race Car Vehicle Dynamics*, convinced the Four Wheel Drive Auto company to let him test the car. Bill ran the car at many races, ranging from Pike's Peak to the Watkins Glenn Grand Prix, Mt. Equinox (where he won and set a record), and various other Sports Car Club of America (SCCA) events until 1951 when he returned the car to the FWD Museum. The car stayed in the museum until the mid-'80s when it was purchased by noted historian and collector Buck Boudeman. Buck sold the FWD to another noted historian and collector, Robert (Bob) Rubin in the early '90s. Bob engaged Chris Leydon's outstanding restoration shop to restore the car, and at about the same time, Griff began to write this book. In the midst of the restoration the car was sold to my brother, Dean Butler, who had the car prepared to full racing condition. Since 1997, the car has been raced numerous times in both the United States and Europe. The car's home is in Cincinnati, Ohio, at Zakira's Garage, which does maintenance and race preparation work.

Buck, Bob, and Dean clearly see themselves as custodians of this unique race car. Their perspectives on the car are interesting and best told in their own words.

From Buck:

I have loved cars all my life. I studied Automotive Machine Shop in College, worked for Holman-Moody in Charlotte, North Carolina and later for the Gilmore Car Museum. There I gained experience with Bugatti, Hispano-Suzia, Isotta-Fraschini, Duesenberg, Packard, Losier, Stanley, White, American Underslung, Delaunay-Belleville, and others.

This brought me closer to some of the great designers. I began to look at these cars as one would look at a great painting and the designers became the artists.

I decided that I liked cars that related strongly to the individuals that conceived them; did the engineering, drafting, pattern work, castings, fabricating, machine work, assembly testing; and, when they were done, they raced them with other artist/designer/builders.

I started to seek out designers that fit my field of interest, Bugatti, Duesenberg, Miller, Stanley. I tried to acquire examples of their work. I studied and compared them.

Eventually I decided that Harry Miller struck a chord in me. I spent my spare time learning what I could about him and his work. Miller came from a very self-reliant and artistic family. His father was an accomplished artist and musician. Miller expressed his artistic aptitude through the automobile much like Bugatti.

As I started to get more familiar with Miller's work, I felt like I knew him better. Each time I found a Miller part I would try to understand it. Why did he do things the way he did? They were like a window into his mind. I tried to see the world as he saw it. Each time I found something Miller, it seemed like I put one more piece of the Miller puzzle together.

I acquired a copy of Borgensen's *The Golden Age of American Racing*. I was stunned! Here were all my heroes!

I kept tracking down leads and eventually I discovered there was a Miller still in the possession of the original owner, the F.W.D. Truck Company in Clintonville, Wisconsin.

I made it my goal to acquire this intact unaltered car. Over a period of about ten years I called repeatedly, asking if the Miller "could be bought." I always got the same answer "son if you want that car you'll have to buy the company." Then I called and talked to the new President. He told me they had merged with Segrave Fire Truck Company and they would like to have some restored fire trucks in the museum. That was what I was waiting for! After some on/off negotiations I bought the car!

I was delighted to find all the paperwork on the car intact. Here was the complete package: the car, spares, tools, drawings, correspondence between Miller and FWD from day one till they quit racing!

I am a dreamer but I didn't even dare dream that one.

I brought the car home and went to school! I studied it, cleaned it, I read all the paperwork over and over until I understood it.

I got the car running and drove it. WOW, this car is remarkable!

I had to share this find. I took the car to Winchester, it ran great.

Several people whose opinion I respect drove it, Bob Sutherland, Amil Andres and others. They all agreed this car WAS SPECIAL. The first four-wheel-drive racer that really worked!

I satisfied myself that I was up to speed on Miller FWD and my quest moved me in new directions.

Bob Rubin appeared on the scene, we visited. He was very knowledgeable and appreciated the history of this car. It was time to share. I sold the car to Bob. After a time Bob sold the FWD to Dean Butler, another worthy owner. Dean has restored the car and vintage raced it all over the world.

I will always remember this car with great fondness, it brought me closer to Miller.

Artists die, but art lives on forever!

From Bob:

I am a great believer in the notion that life is what happens to you while you are making other plans. In this spirit, I discovered Harry Miller quite by accident, while pursuing a deliberate, even relentless, quest for great Ferraris. The Harrah collection was being dispersed, and I wanted what he had. Unfortunately I got there too late for any of the really tasty artifacts of Modena. But when I asked Ken Gooding, the executor, whether he had any interesting and important cars left, he mentioned the front drive 91. All I could say was, "What's a Miller?" My next call, to a fellow Ferrarist whose horizons were already broader than mine, told me that if I could have that car I should just buy it—which I did.

That acquisition was the key to one of the great archaeological and spiritual adventures of my life. It led me to Griff Borgeson, a mentor who is sorely missed. I also met the hard-core Millerite community—Buck Boudeman, Chuck Davis, Bob Sutherland, Dave Uilhlein, and, of course, the late Mark Dees. And I like to think I helped bring Harry Miller into the mainstream of automotive history, where he truly belongs—adding dozens if not hundreds of people to the core of true believers, plus putting Harry up there with Bugatti, Ferrari, and damn few others in the pantheon. The front drive 91 now lives in the Smithsonian, and Griff's book on Miller continues to sell like hotcakes. And who can forget Miller Year at Monterey?

The 91 was a legendary car, its buzz helped, ironically enough, by the greater international stature of one of its owners, Ettore Bugatti, who plagiarized much from it. The FWD, on the other hand, was more complicated to understand, but when I saw it for the first time at Buck's, I was immediately transfixed. The French have a wonderful expression for something that is untouched: "dans son jus," loosely translated, "in its juices." I beheld the FWD in its juices. It was not difficult to convince myself that further financial overextension was in order, and the car was soon mine. Eventually I sold it to Dean Butler, who by running it in European vintage races is doing God's work to put Miller's reputation where it ought to be. To be in a chain of custodianship bracketed by Buck and Dean is something to be proud of.

The documentation that exists with the car, which forms the backbone of this book, is nothing short of extraordinary. Except perhaps in the case of certain Mercedes factory racers whose dossiers are still in the hands of Stuttgart, no 70-year-old race car has this kind of archive intact. The mixture of Griff's historical genius and this rich primary material is a combustible one indeed.

From Dean:

Old race cars are interesting to many of us, but they come truly "alive" only when we know the personalities, achievements, failures, strengths, weakness and general history of the people who conceived the cars, designed them, machined the parts, assembled the cars, worried over them, drove them and repaired them—the people who made them "live." Griff Borgeson, perhaps more than any other motor racing historian, understood all this and breathed unique historical life into his writings. Many of Borgeson's books are among the best we have about a given make (Alfa Romeo and Bugatti), genre (early American racing cars), or subject (history of the twin cam engine). Borgeson's great classic, *The Golden Age of the American Racing Car* makes American racing cars of the teens and twenties live and breath like no other author has ever achieved. Now, in *The Last Great Miller*, Borgeson has brought life to an historic, unique model of race car in his special manner. I know of no other race car history which comes so alive as the FWD Miller.

As current custodian of the remaining FWD Miller, I am committed to keeping the car itself "alive" by keeping it in racing trim and by demonstrating the car as often as possible. It is important that current and new generations know what these cars look, sound, and smell like in their racing environments. And, driving the FWD Miller on a race track is, of course, a lot of fun. However, the fun was immeasurably altered for the better after I read Borgeson's first draft (which, fortunately, I was able to discuss with him prior to his death). Now, when racing the car, I know enough about the man/car relationships of Miller, Goosen, Offenhauser, Brisko, Rose, and Milliken to have a far deeper appreciation of the feel of the car, its cornering, braking, and general handling. I also like to believe that I really am uniquely able to "be" Brisko, Rose or Milliken at the wheel. It's wonderful.

Thank you, Griff—and you will always live in my memory, silver pony tail and blue jacket blowing in the wind as our great friend, the irrepressible, forever young Bill Milliken drove you around Bridgehampton at speed in the FWD Miller.

My wife, Dolores and I have had the pleasure of acting to move Griff's work through to publication. This project has given us the opportunity to interact with individuals connected with the FWD Special. Each of these individuals has not only helped us move the project along, but has deepened our appreciation for the genius of Harry Miller and his contributions to racing. My sincere gratitude to Dolores, whose tireless effort and detective work have knitted together Griff's text, documents, and illustrations.

Donald D. Butler
Cincinnati, Ohio, June 1999

Acknowledgements

We who have had the pleasure of creating this modest volume wish to gratefully acknowledge the following individuals for their valuable inputs:

Jim Bibb, Buck Boudeman, David Brownell, Roland Bugatti, William A.M. Burden, E. Dean Butler, Don Chew, Steven J. Church, James D. Crank, Charles F. Davis, William Digney, René Dreyfus, Robert Fabris, Nick Georgano, Alec Giaimo, Leo Goossen, George and Stacey Hanley, Jim Himmelsbach, Hayashi Kageyama, Shotaro Kobayashi, William J. Lewis, Chris Leydon, Rita Leydon, Karl E. Ludvigsen, J. Alfred MacKenzie, Doug Milliken, Bill Milliken, Overton A. Phillips, Walter Pittoni, André Rheault, William Riley, Laurent Rondoni, Michael D. Rosen, Robert Rubin, John M. Russell, Peter Shaw, James Simon, James Standberg, William A. Turner, David Uihlein, Frederick A. Usher, Peter D. Williamson, and Mike Yust.

Photo Credits

Credits are given to all photos and figures at the end of the caption. Abbreviations used in photo credits are indicated below in parentheses.

Griffith Borgeson Collection (GBC)
William A.M. Burden
Carlo Otto Brambilla
Charles F. Davis
Bob Dunsmore Photography
Four Wheel Drive Auto Company (FWDCO)
Leo Goossen
Frank Gudaitis
Steve Havelock
Indianapolis Motor Speedway Corp. (IMS)
Kirchoff
William F. Milliken Collection (WFMC)
Ernie Olson
Packard Motor Car Co.
Pete De Paolo (PDeP)
Robert M. Rubin Collection (RMRC)
David Uihlein
Frederick A. Usher (FAU)
Frederick A. Usher Collection (FAUC)
Michael Yust

Chapter 1

Wisconsin Über Alles

"There may have been a four-wheel drive prototype or two in the world before, but it was when the four-wheel drive trucks came to France by the thousands during the First World War that we Europeans learned what four-wheel drive could be. They would go anywhere, with amazing reliability, and they came to symbolize those qualities. We loved them." These words came from Paul Berliet, one of the world's great authorities on the history of industrial vehicles.

The Four Wheel Drive Auto Company was founded in Clintonville, Wisconsin, in 1909, based on patents and experimental vehicles that dated back to 1907. Its founding president was a young local attorney named Walter A. Olen. The firm swiftly became the world

THE NEW AUTO A DANDY

Zachow & Besserdich Put Their Four Wheel Drive To A Hard Practical Test

On several occasions, we have told of the newly patented four wheel drive automobile Zachow & Besserdich, our local machinists, had patented, and throughout the past summer, many have watched its construction and have been eagerly watching for the machine to be put to a practical test.

The test was delayed several weeks because of a poor boiler which was shoved onto the firm by an unscrupulous firm at the outset, and this had to be removed, thrown away and an entirely different make of boiler installed before the auto was ready to be tried out.

With all connections made, everything was in readiness last Saturday, Dec. 26, 1908, and since that time, the inventors could have been seen almost any day, trying the merits of the car. With an ordinary dry goods box placed on the frame work of the auto to answer the purpose of a seat, they have plowed through snowdrifts three feet deep, pushed huge billows ahead of them, gone over the roughest and most slippery roads and climbed the steepest hills to be found as well as to speed the machine up to a 25 mile hour clip and through it all, the newly patented mechanism of their four wheel drive has answered perfectly every demand made upon it.

There has been no digging in of the rear wheels as is the case with the ordinary auto which is driven from these 2 wheels alone, the apply of power to all four wheels keeps it from skidding, makes the travel over the surface and with so much ease that no obstacle seems too great for it to surmount.

The new driving mechanism has been so carefully and perfectly planned that it does not in the least interfere with the steering or handling of the car nor detract from its beauty and unless some defect shows up later on, which is practically out of the question, it will revolutionize automobile building.

Those who were the closest observers of the test cannot say enough in praise of Zachow & Besserdich's new four wheel drive auto and there is considerable talk of getting the inventors to organize a stock company and manufacture the autos in Clintonville. There is only one change contemplated by the builders and that is to make a four cylinder gasoline instead of a steam car, as they believe the former will be less complicated and much handier than steam.

Illustration 1-1
1909 newspaper article on four-wheel drive.

leader in its field and when, in 1931, Harry Miller sensed that the time had come for the application of four-wheel drive to both racing and passenger cars, his thoughts turned automatically, and with pleasure, to Clintonville, which was not far from his place of birth, Menomonie, and was in a region that he loved well. Walter Olen still headed the company, and what he may or may not have confided to Harry Miller was that the maiden names of his locally born mother, Amelia, and of his wife, Cora, both had been Miller. In any case, this haven for Teutonic immigrants was a small world.

Following yet another Miller victory in the Indianapolis 500 on May 31, 1931, Harry Miller took advantage of his proximity to Clintonville and made the 300-mile run to the northwest. His traveling companion in his L-29 Cord Brougham was an old and close friend, Harry Hartz, whose front-wheel-drive Miller racing car had just finished second at the Speedway and would win the big race the following year. Harry Miller was at the height of his career as master builder of racing cars in America and he was received with fitting deference by Olen and his executive staff. The purpose of his visit was to present to the company's board of directors his vision of four-wheel drive not merely in trucks, but in both road and racing cars, and starting immediately. Harry Miller could be eloquent with listeners who spoke the language of machinery, and the board was impressed.

The key to four-wheel drive and, in the case of more than two live axles, all-wheel drive, is front-wheel drive. That is to say, it is the acceptable transmission of power through those wheels of a vehicle that must do its steering. At low cornering speeds, such as those within the performance limits of early trucks, the jerkiness inherent in the simple and inexpensive cardan-type universal joint was acceptable. In the immortal words of a French pioneer, "C'est brutal, mais la marche." [It is brutal, but it works.] As cornering speed increases, however, the jerkiness of power transmission through an angle becomes increasingly hard on driver and on mechanism. Miller and his collaborators were confronted by this problem in 1923 and, according to Leo Goossen, their first solutions employed simple cardan joints. Negative feedback from drivers was immediate. It took the design team a good deal of pioneering to develop an approach to an acceptable constant-velocity U-joint, that is, one that did not buck throughout a useful range of deflection of its input and output shafts; and one in which both shafts rotated at the same angular velocity. The Miller solution to this problem worked wonderfully well, as the racing record shows. From the moment of finishing second in its maiden gallop at Indy in 1925, Miller front-wheel drive electrified technical and popular imagination everywhere. Its creator, who lived in the future, undoubtedly knew then that he possessed the chief key, not only to the four-wheel-drive competition machine, but to the four-wheel-drive super road car as well. Their time was not yet, but it would come.

The matchless virtue of basic four/all-wheel drive is that it distributes engine output between each of the vehicle's wheels. Thus, in cases where the contact patches or footprints of a pair of tires are insufficient to transmit a given amount of power/torque to irregular or slippery surfaces, all-wheel drive at least increases that capacity. This, of course, was the great trump card of the Four Wheel Drive Auto Company, and it was accompanied by side benefits, such as reduced stress and wear of mechanical parts and tires. Among commercial vehicles, all-wheel drive was king on mud, sand, snow, and ice, as well as on broken terrain.

In the specialized world of automobile racing, techniques had always existed for the use of controlled skidding while cornering. Breaking traction while in top gear on straightaways

Illustration 1-2 A 1917 advertisement explains the anatomy of the Four Wheel Drive Auto Company's go-anywhere products.

did not become a problem until the advent of the elevated power outputs of the late 1920s. Power outputs became so high that very minor surface irregularities could induce sudden wheelspin and loss of traction. One of the reasons that eight-cylinder engines were so popular was that their power impulses were half as abrupt as those of equivalent fours. Toward the end of the supercharged 91 cu. in. formula of 1926–29, conditions reached a point where top cars were spinning their wheels at top speed. It was for this precise reason that Miller had gone to work on the development of a V-16 91, to further reduce power-impulse impact.

Miller was probably already working on the alternative of distributing the impact of power impulses among four wheels, instead of only two, when the Crash of October 1929 transformed the economic environment of the planet. The luxurious sport that was motor racing received the full brunt of the immediate withdrawal of capital. All thought of maintaining, in operating condition, the board speedways, which were the diadem of American automobile racing, was immediately abandoned. With the exception of the hard-surfaced Indianapolis Speedway, racing moved to dirt tracks, which were inexpensive to build and maintain. Moreover, most of them already existed as one-mile horse tracks at major fairgrounds. Aside from Indianapolis, the annual national championship of the American Automobile Association would be contested almost exclusively on dirt ovals. On all of them, traction would be a major problem, on the chutes as well as in the turns. Miller recognized the arrival of the age of four-wheel drive.

Another new development had a place in this new reality and it, incongruously, also coincided with the arrival of the Great Depression. It began with the introduction of the eight-cylinder Model J Duesenberg in 1929, followed swiftly by Cadillac, Franklin, Lincoln, Packard, and Pierce-Arrow V-12s; and Cadillac and Marmon V-16s. All were super-powerful and could put to use, but lacked, the benefits of four-wheel drive. Miller spelled out the historic concatenation of events to Olen, concluding with the company's ideal position to profit from it. Olen was moved and told Miller to put his ideas in writing, which he would present to his board at its upcoming meeting in a few weeks. The letter, which follows, is the earliest known document pertaining to the history of Miller four-wheel drive.

July 3rd, 1931

Mr. Walter A. Olen,
Four Wheel Drive Auto Co.,
Clintonville, Wisc.

Dear Friend;

To start out with I want to thank you for the courtesies which you extended to Mr. Hartz and myself while we were there.

I will give you some quotations on the cars that we were talking about. First, the Four Wheel Drive car for racing purposes with one of my 4-cylinder engines complete would be $10,000.00; the same chassis with one of my new 8-cylinder engines would be $15,000.00. I have never sold a Front Wheel Drive car for less than $15,000.00, so you can see that this is a very fair price. The reason for making this price is that I want very badly to build this Four Wheel drive racing car for Indianapolis.

In regard to the experimental Four Wheel Drive super-touring car the cost of same would be approximately $25,000.00, as you understand patterns, jigs, tools would all have to be made as the engine would be a 16-cylinder engine of complete new design. Also, spring suspension, chassis, radiator and body design would have to be something entirely new which would require quite a bit of work in the engineering and drafting rooms.

This price you realize is certainly a minimum for a job of this type.

Of course, this car would have to be exceptional in looks. The riding qualities would have to be superior to any known car made possible by our new spring suspension. The engine would have to be a beautiful job of enough horsepower so that the car would do more miles per hour than any present day stock car.

I would like to start work on these cars as soon as possible, especially on the race jobs, so in the event there were anu bugs in it, we would have plenty of time to work that out. While this work was being done, the designing of the other car would naturally be in progress.

Leon Duray, who is now associated with me and who is holder of many existing world's records both here and in Europe, most all of which he established in one of my Front Wheel drive cars, one of which is the world's circular track record of 148.02 miles per hour on the Packard Proving Grounds in Detroit with a 91 cubic inch engine which is quite a performance due to the fact that in breaking this record, he defeated all size engines since the beginning of time. He also holds the fastest single lap at Indianapolis and the fastest qualifying time at Indianapolis, together with many ocher world's records in this country and Europe. He owns and drives a new model Duesenberg, which chassis originally sold for $8500 and they have since raised their price to $9500 and as near as I can ascertain, they are turning out a car per day.

If you would be interested, we could bring Mr. Duray's car with us for your inpscetion when we come.

I am enclosing a photograph of my 16-cylinder engine which I had in a car at the recent Indidnapaolis race, also enclosing two views of a special job I made for Phillip M. Chancellor, which was powered with a V-8 engine, especially designed for the car. This car did 110 m.p.h. in second gear on demonstration, and of course will run 130–135 mph in high gear.

I am leaving for Detroit within the next few days, and if I remember correctly you were to have your meeting the first or second Tuesday of this month. If I could be of any assistance to you, please let me know and I will make a trip to your place as soon as I arrive in Detroit if you so desire.

As I will not be here to receive an answer to this letter, please make your reply to 2843 E. Grand Blvd, Detroit.

Hoping to receive a reply to this letter upon my arrival in Detroit, I remain

Very truly yours,

Harry A. Miller

Harry A. Miller
HAM

Lacking a stenographer's initials, this document has the air of having been typed by Miller himself and of representing his own style of expression. Among other things, it tells us that:

1. Earlier discussion with Olen had dealt with the construction of a team of cars, not just one.
2. The global plan included the construction of a prototype four-wheel drive "super-touring car" that would surpass any existing production car.
3. The target to be surpassed was the Model J Duesenberg.
4. The supercar was to have a V-16 engine, such as the one in Miller's personal L-29 Cord, which he had exhibited at Indianapolis a fcw weeks before.
5. The proposed supercar would be of the level of quality of the superb one-off road machine that Miller recently had created for Phillip Chancellor.

6. Within the global plan, priority of construction would go to the real standard-bearers, the competition cars.
7. This was not considered to be an opportune time for Harry Miller to reveal his designs on the World Land Speed Record, using four-wheel drive and engines of his own conception.

Miller had a clear-cut plan for building the ultimate in American racing and passenger cars. They would have in common the unique distinction and advantages of four-wheel drive. What more logical and ideal backer than the eminently reputable company which for decades had pioneered the principle and owned the name?

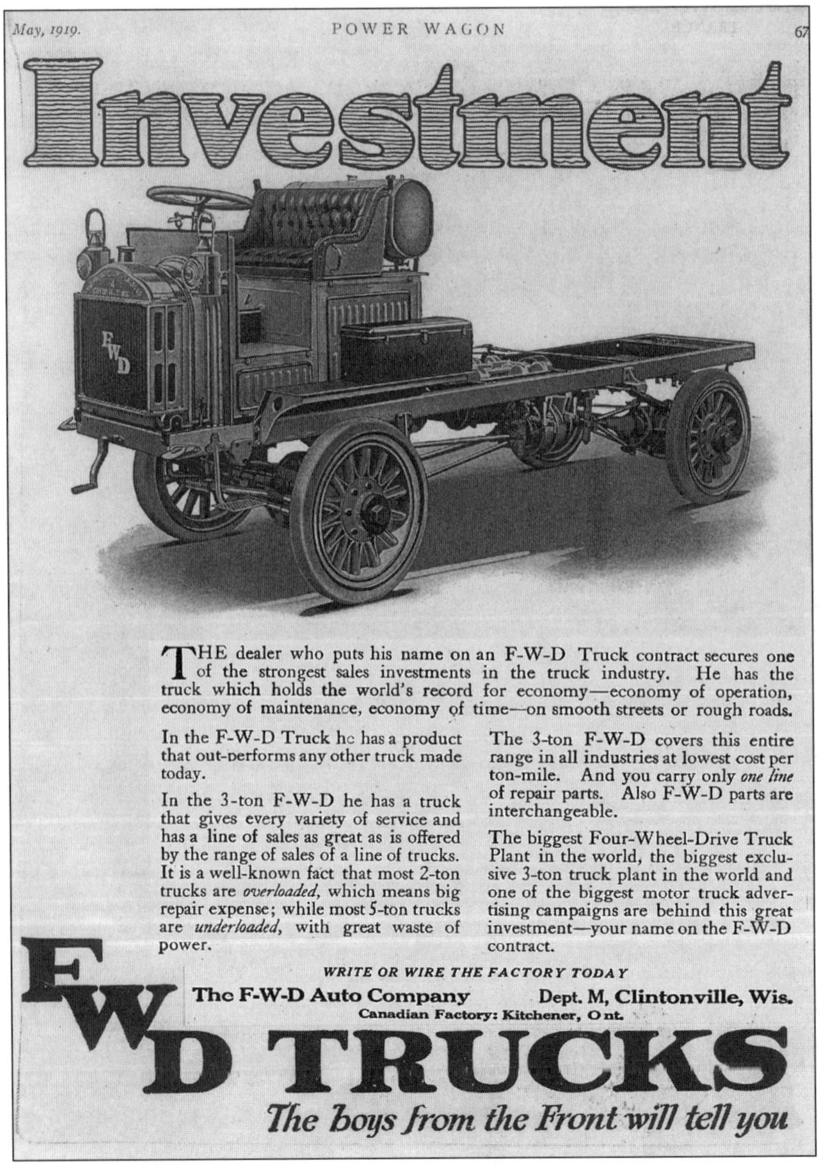

Illustration 1-3 A 1919 ad, featuring a 3-ton bare chassis, stresses solid value. Legendary performance in World War I is summed up in "The boys from the Front will tell you."

Chapter 2

The Historical Setting

Whether or not Miller was aware of the Four Wheel Drive Auto Company's financial situation, the timing for putting his plan into action was perfect in some ways, disastrous in others. As World War I had done for countless industrial activities, it was this conflict that put the company and the four-wheel drive principal seriously in business, and profitably so. When peace finally arrived, late in 1918, international markets were glutted with war-surplus equipment of all sorts, which helped to create a series of very lean years. In the case of the Four Wheel Drive Auto Company, it was not until 1928 that the firm began to operate in the black again. Profit for that year was a promising $109,000. The next year brought a plus yield of $213,000, but also the famous Crash of '29, which became the Depression of the '30s. The speed with which money went out of circulation was astonishing. The Four Wheel Drive Auto Company's sales plummeted with the rest, but the firm, with its coffers well stocked, faced the crisis head-on with militant sales and advertising campaigns. Its directors voted a budget of $100,000 for publicity in 1930, a fact that the trade journals hailed.

In 1931, the year in which Miller made his proposal to it, the company had reason to spend even more on advertising. The company was celebrating its 20th year of existence, and the construction of its 25,000th vehicle. Still, sales for that year declined much more, resulting in operating losses that would worsen until another war came along. Thus, if Olen pinched the pennies until they screamed as he watched Clintonville's banks close their doors, it was because saving his company and the jobs of hundreds of employees was more important than gambling on the race track. One crazy fling was enough, as a one-shot advertising novelty.

Harry Miller's financial environment was more moribund than most because, as soon as money got tight, the bottom dropped out of racing as a livelihood. Fred Offenhauser told me something like, "During those years we managed to stay alive trading nickels and dimes, hoping that someday a half-dollar would show up." Miller's genius had helped to make the Roaring Twenties what they were, and they rewarded him well, both materially and with recognition. His client/patron, Errett Lobban Cord, was one of the few who had a clear vision of the coming Crash, and he urged Miller to cash in his chips while he still was winning. Miller had a standing offer of $150,000 for his business from a group of operators headed by one George Schofield. In agony of spirit on February 7, 1929, Miller put his name on the line, took the money, and walked away from the plant and staff that had been his life. It had been a quite literally paternalistic environment, a sort of freemasonry of the machine in which he was known to the initiates as "Father." The idea was that he would retire, and he and wife Edna and son Ted went east, putting the West Coast behind them. That lasted about a year, when Harry moved back to Los Angeles. He opened a small machine shop early in 1930 and began puttering with projects that had been gnawing at him ever since he had turned his back on his world. That December, the company that called itself Miller-Schofield Motors Incorporated joined the swelling bankruptcy rolls of the country. Miller, more aware than anyone of the inestimable value of the unique talent pool that had been drawn to him over the long years, hired it back, as though the Depression would go away at any moment. In addition to being an artist rather than a businessman, Miller depended heavily for guidance on what he called "a spiritual control." He recovered all of his patterns and drawings from defunct Miller-Schofield, purchased a fine, large building at 6233 South Gramercy Place, near Western Avenue and 63rd Street, poured money into outfitting it, and prepared to pick up where he had left off as America's premier racing-car builder. On January 11, 1931, the Los Angeles Times announced "The Most Powerful Passenger Car Ever Built," on display in the new plant, along with the under-construction V-16, which Shorty Cantlon would drive at Indy on May 30.

Miller did not intend for these cars to be one-offs. He saw them as the first-offs of road and competition machines that would be built in limited but still-profitable series, as his recent 91s had been. Mark Dees [Miller historian] tells us that there was a plan for Wade Morton*, a long-time key man in the Auburn-Cord-Duesenberg realm, to form a new corporation for the marketing of an L-29 Cord-based supercar. The Bill White V-16, with its De Dion suspension all around, was conceived to race on dirt or pavement, powered by either the V-16 or a less-expensive, compact, new Miller four-cam V-8. Such cars were conceived to fill a new need, but their potential buyers needed money more badly than they needed the cars. The same was true of the four-wheel-drive racing cars and their four-wheel-drive super gran turismo sisters. No one could or would spend the money to debug and develop the prototypes. The situation was exactly that of the original front-wheel drive 122 when it first took to the bricks and boards. If the funds had not been available (from Cliff Durant) for the required debugging, the birth of modern front-wheel drive would have been delayed indefinitely. When Miller built a second four-wheel-drive racer, allegedly on speculation, he did not see his investment that way. He saw himself standing firmly on the threshold of a breakthrough that was just as fundamental as that of front-wheel drive. He was right, but no money was coming in, and his was running out. When the paltry sums that cured the problems of early Miller 4WD finally were spent, history had passed it by. And history did not look back when those who had believed in the principle all along were proved right.

As we read the pages that follow, let us never lose sight of the fact that we are not dealing with a few isolated experiments, but with clairvoyant anticipation of engineering reality to come. This is what Harry Miller's life and work were all about.

* For much information about Wade Morton see the author's book, *Errett Lobban Cord,* Automobile Quarterly Publications, Princeton, N.J., 1983.

Chapter 3

Currents of Technical Change: Kingpin Confusion and Compromises

Late in 1928, and shortly before the sale to Schofield, a Santa Barbara millionaire commissioned Miller to create "the finest and fastest sport roadster ever built, without regard to cost." The car was to be a gift for the man's son, Phillip Chancellor, apparently to mark his 21st birthday. Delivery was to take place in March 1930. This was a dream assignment and, Goossen told me, Miller rose to the occasion with a design that was entirely new from stem to stern. It benefited from the team's now considerable experience with front-wheel drive, not only in racing applications, but also in the L-29 Cord passenger car. All of that experience was marked by weight distribution that did not put sufficient load on the tractive front wheels. Miller sought to eradicate that defect, starting with a new concept in transaxles whereby the change-speed gears, which were arranged longitudinally, extended ahead of the front-wheel centerline. The whole front-drive unit was mounted very far forward, with the De Dion tube passing under it to allow this. And this time the change-speed gears were located properly, on the low-torque side of the final drive pinion, permitting easy shifting. (Goossen termed the mechanical solution "ideal.") Then, bolted to this assembly was a short, compact V-8, which added its weight to the loading of the front wheels. Block-wise it was half of a Miller marine 16C/620, thus a 310.

The car was still in the layout/drawing stage when Miller headed east in February of 1929. Although Schofield would continue to profit from Harry's name and fame by listing him as "director-general of engineering," Miller had little to do with such matters. The actual detail layout of the Chancellor car was done by Goossen, with the help of a nephew of Miller's named Walter Steele. When drawings were sent to Harry for review they were a very long time in ever coming back. But the car got built by his old team anyway, and he did present himself at the unveiling of what the press hailed as his masterpiece. Whether its delivered price was $30,000 or only $25,000, both of which were reported, it was one of the few most expensive cars in the world.

The general layout of the new car's chassis resembled closely that of its oval-track forebears. It used simple channel-section frame rails, with kickups over the rear axle. Front suspension was by over-and-under quarter-elliptic springs, with their eyes at the front. The rear axle rode on outrigger half-elliptics that were underslung. Caster, camber, and toe-in of the front wheels were not adjustable; they were designed into the steering knuckles, the kingpin forks, and the front springs. This basic architecture had proved itself over tens of thousands of miles of high-speed driving, but when the younger Chancellor took his first spin in the car he found it to be undriveable. He and his father were outraged and threatened to annul their contract with Miller. Whoever was speaking for Schofield management promised to rectify the situation and did, although Chancellor seemed never to trust the car, and soon sold it.

The story became a part of southern California legend. David Gray of Santa Barbara was a friend of Chancellor and rode with him on the car's maiden voyage. As he was said to have told the story, they were not even going fast, when the front wheels began to shimmy and wobble terribly. Stopping, they eventually learned that the wheel hubs, with their right- and left-hand threads, had been interchanged by error, and the Rudge hubcaps had begun to unscrew themselves. A not very likely story.

A very different one was related by Bunny Phillips to Frederick A. Usher in 1984, in the course of a series of interviews devoted to the career of the great Miller and Bugatti specialist. He told Usher, speaking of the Miller organization:

"They didn't know much about front-wheel drive at the time of the Chancellor job. Instead of having positive caster, with kingpin tilted back to make them easy to steer they gave them a negative caster. The cars were all right as long as they were going fast, but at low speeds they would shake. They would develop this tremendous shimmy and you couldn't hold it. It would just go where it wanted to. That was the fault of all the early front-wheel drive cars. They all had that negative caster."

According to Bunny, the Chancellor car steered itself into a stout roadside hedge. He added that the Schofield people, in an effort to make their client less unhappy, arranged with Leon Duray to give him, as a loaner while his car was being fixed, one of the Bugatti Type 43 Grand Sports for which Duray had swapped the Packard Cable Specials at Monza the previous September.

Phillips's negative caster theory is not confirmed by existing drawings. Speaking of Miller front-wheel drive in general, Leo Goossen told me:

"The kingpin and U-joint were smack in the center of the wheel, so you had a pivot when the wheel was turned. The centerline of tire and kingpin intersected at the center of the tire print. This gave a minimum of problems and worked out very well. There was no sliding of the tire to cause wear."

"The front springs were mounted at a four-degree angle, the caster angle. They were positioned so that the wheel-center would follow a straight vertical line. There was about one-half a degree of camber to the wheel."

"We made and machined the front-wheel drive front hubs and they were laced up [with wire spokes] by Firestone, who made the rims. The rear wire-wheel hubs were stock Firestones. We made the inner hubs [the male hubs onto which the wheels fitted], front and rear."

Because of the large size of whatever they used for front outer U-joints, "[w]e had to make those huge front hubs [inner and outer]. Later, when Miller made the four-wheel drives, he slanted the knuckle so that he could use the same stock shell front and rear. He had to get that knuckle out of the way to do it."

This statement asserts that the relatively ideal geometry of classical Miller front-wheel drive was sacrificed in favor of wheel interchangeability. One immediately checks photos of the Chancellor and Burden cars, sees that their wheels were interchangeable, and seeks to draw conclusions. And the L-29 Cord, which was designed at Miller's before the Chancellor car, too, of course, had interchangeable wheels, forcing its kingpin axes well inboard. And it was a bear to steer. Still, we know from Milliken, Boudeman, and Rubin that the four-wheel drive FWD Special is a fine car to drive. What was done to make it so?

The FWD Miller has outboard brakes that could lead to "brake-pull" if the kingpin intersections with the ground were other than close to the center of the tire point, i.e., "center-point" steering. Miller had experience with front wheel drive cars with both inboard and outboard brakes, so he must have been aware of the problem of "brake-pull." On the FWD Miller, the kingpin appears to be located for center-point steering. Driving experience tends to confirm this: There is no noticeable brake pull or shaking of the steering on rough surfaces. The centering action of the steering is positive, but small, as it comes from the tire aligning torques only and not from castor action. Steering forces are low but acceptable.

Respiration by Diktat

The engines made by Miller following his year's absence from Los Angeles differed from their predecessors in some important ways. The most significant differences were in his induction systems.

Miller's early, normally aspirated straight-8s, with a separate respiratory tract for each cylinder, constituted a close approach to engineering perfection in that sector. With the advent of forced induction this emphasis upon super-efficient gas flow lost its urgency. But with what may have been his last 91 engine, that of the Hepburn car of 1928, Miller proudly returned to separate ducting for each port, even though a husky centrifugal blower was there to pile-drive the mixture to its destination.

When, in late 1929, an ex-Miller ace, practical engineer Riley Brett, decided to build a V-16 engine based on Miller 91 practice and components, he had the idea of having the inlet ports rise vertically through the cylinder head, instead of reclining horizontally in the habitual manner. This stratagem lent itself ideally to the newly developed technique of downdraft carburetion, with the added advantage of holding engine width at a minimum. Rules for the 1930 Indianapolis race were warped grotesquely in order to stimulate the use of stock-block engines, among other vulgarizations of the discipline. Only two carburetors could be used, and superchargers were banned.

Each bank of cylinders of Brett's engine for the Sampson Special had eight nice, separate inlet ports. To get the fuel/air mixture to them from a pathetic, centrally located downdraft carburetor, Riley created a labyrinth of manifolds that was a respiratory horror. Nevertheless, that lashup permitted driver Louis Meyer to finish a handsome 4th in the 1930 Indianapolis Sweepstakes.

Much of Miller's year of leisure during this period had been spent in seeing America, from the controls of a personal L-29 Cord. This gave him more than ample opportunity to analyze the defects of this product and to reflect upon how a proper front-wheel drive supercar should be made. Thus, his first project upon returning to the City of the Angels was to translate these ideas into metal. All of the Cord's objectionable features were in its powertrain. Goossen was not at liberty, and Miller put engineering draftsman Walter Steele to work on laying out the engine. He assigned the transaxle to one S.C. Chesterfield. This front-drive assembly probably incorporated the "ideal" design features developed for the Chancellor car, with still further improvement.

The engine for the supercar was a V-16 whose 2.63 × 3.5 bore and stroke gave it a displacement of 303 cu. in. This was well within the new Indy displacement regulations, should one elect to produce a racing version of the engine. To keep frontal area at a minimum, Miller chose an included angle of just 45 degrees for the cylinder banks. Side-by-side connecting rods, tubular as always, were used, requiring that the banks be staggered relative to each other. In this passenger-car version of the engine, its four overhead camshafts were driven in relative silence by connecting rods and eccentrics, instead of by traditional singing, screaming spur gears. The principle was not original with Miller, but this interpretation of it was uniquely exotic.

Following normal Miller practice, each bank of cylinders was made up of two four-barrel blocks with integral heads. With Brett's approval, Miller endowed them with downdraft inlet ports. Taking this idea a long step further, he eliminated external manifolding by making it an internal element of the head/block casting. And, instead of ideally separate ports for each cylinder, Miller chose to use a Y-shaped passage for each pair of cylinders. These are the famous "siamesed" ports. Miller did not choose to apply to downdraft carburetion the separate-tract principle that he had pioneered with updrafts a decade or more before. Instead, he used a single-throat carburetor for each two cylinders, its float bowl occupying longitudinal space that might have been used for much-improved breathing.

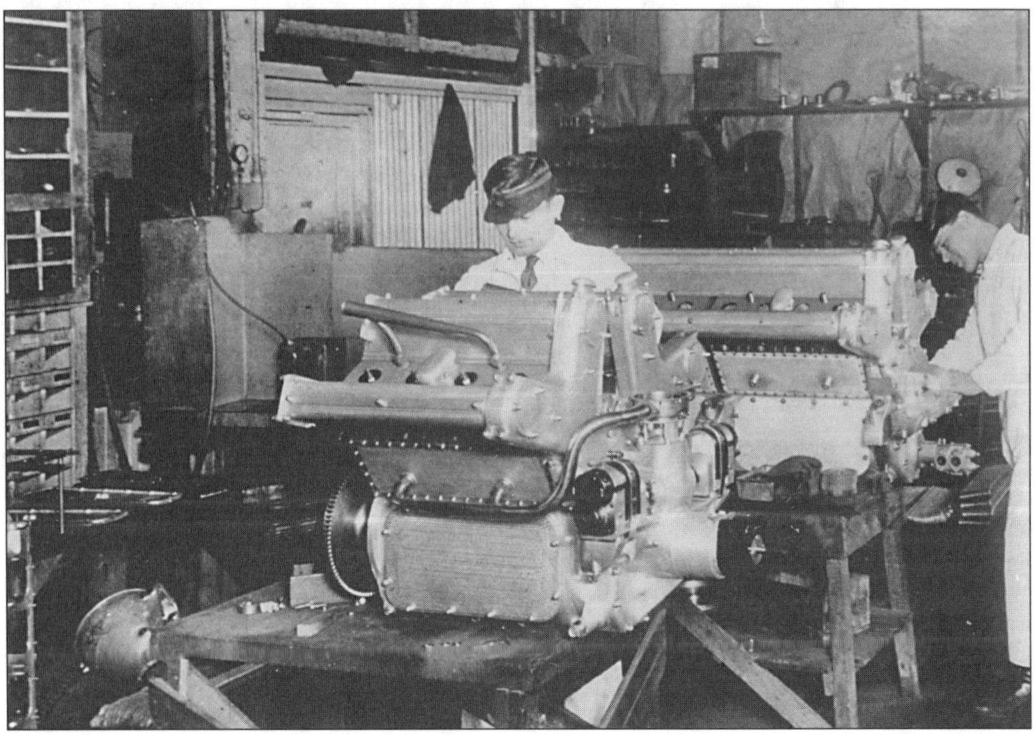

Illustration 3-1 Miller conceived the 90-degree 310-cu. in. V-8 engine for the Chancellor car with concentration of weight on the front wheels foremost in mind. A marine V-16 is being assembled in the background (GBC ex Kirchoff)

Illustration 3-2 The Chancellor chassis was a consummate gem. It had double drag-link steering, as can be seen here. This feature dated back to Harry Miller's Golden Submarine. (GBC ex Kirchoff)

Illustration 3-3 Frontal view of Chancellor car shows structural details that are identical to those of the classic Miller four-wheel-drive competition car, but engine and transaxle are farther forward. Note the stiff, bent-back De Dion tube forks. (GBC ex Kirchoff)

Illustration 3-4 Frontal view of Chancellor car, ready for the road. The bumper is still a bit askew. (GBC ex Kirchoff)

Illustration 3-5 Running boards were unnecessary on this very low car. Identical hubs made all road wheels interchangeable. (GBC ex Kirchoff)

Illustration 3-6 Body with disappearing top was designed by Gerald Kirchoff, formerly of Murphy of Pasadena. (RMRC)

Illustration 3-7 With top fully concealed. (RMRC)

Illustration 3-8 A very young Fred Usher photographed his brother in this car in 1937, parked on a street in Hollywood. Bunny Phillips had swapped the car's engine for a flat-head Ford V-8 and the owner had stepped up to more modish fenders. (FAUC)

Illustration 3-9 With the advent of downdraft carburetion (Winfields here), practical engineer Riley Brett invented the downdraft inlet port. Leo Goossen drew up these modified Miller 91 blocks for the 1930 Sampson Special. Indianapolis Motor Speedway carburetor restrictions impose this atrocious manifolding. (GBC ex Kirchoff)

Illustration 3-10 The 303-cu. in. 45-degree V-16 that Harry Miller fitted to one of his L-29 Cords was laid out by Walter Steel, a nephew. Note downdraft inlet porting, Miller downdraft carburetors. (RMRC)

Illustration 3-11 Photo of Leo Goossen, taken circa 1965. Around the shop it was always said, "Leo can put the needle into you and you'll never know it." (GBC ex Leo Goossen)

This retrograde progress is not explicable by the fact that the engine, at this stage, was intended for street use. When, later in 1930, the full-race Miller 4C200 appeared, it too had siamesed ports, as did the in-line 8C230, and then the 45-degree V8C308 which powered the four-wheel drives. Miller, a brilliant talent in the fields of carburetion and fluid dynamics, knew what he was doing. The explanation for this apparent enigma is contained in the technical regulations that the Contest Board of the American Automobile Association (AAA), in concert with IMS (Indianapolis Motor Speedway) management, devised for the Indianapolis 500. These regs thereupon became largely dominant for the ten or twelve events in which the American Automobile Association's annual National Championship was contested.

Under what the aristocracy of the machine termed the AAA's "Junk Formula" of 1930, and as a decoy for Detroit iron, piston displacement shot from 91 to 336 cu. in., where it would remain through 1937. Four-stroke engines could not be supercharged and could have only two valves per cylinder. Only two carburetors would be tolerated, regardless of the number of cylinders of a given engine. Hence, the martyrdom of the Sampson Special.

Revised regulations for the 1931 Indy 500 were drafted in the fall of 1930. Due to generalized outcry, one single-throat carburetor now became permissible for each pair of cylinders. As in 1930, twin throats with a single float bowl counted as two carbs. These rules also lasted through 1937, after which AAA/IMS regulations got into line with AIACR* international regulations.

Under these constraints, Miller's choice of siamesed ports and a single-throat carburetor for each pair of cylinders was the height of logic. Everyone was up against the same limitations. Not surprisingly, Harry Miller's way of adapting to them was the coolest of all.

The Double De Dion Chassis

More or less concurrently with his V-16 powertrain project, Miller built the prototype for his 1931 Series competition chassis. It harked directly back to his first front-wheel drives of 1925. That front-end architecture had been refined to a high degree over the years, as the techniques and tooling for making it also had been. In the case of these chassis, steel frame rails of deep channel section were chopped off squarely at the front, and they tapered to graceful kickups over the rear axle. Big cast-bronze perches were bolted to the front of the rails, and the over-and-under quarter elliptic springs were bolted to the perches. The De Dion tube, which was the front dead axle, bore integral flanges, to which the spring eyes were bolted. The steering knuckles, which carried the hubs, were king-pinned into the forks at either end of the De Dion tube. Starting with the 1931 season, Miller gave the rear kickups a sort of ducktail upsweep. To the channel sides he bolted the now-familiar over-and-under quarter elliptics, but with their spring eyes to the rear. These supported a rear De Dion tube, while the live half-shafts and the final drive housing were sprung-weight bolted to the frame. This mirror image, "double De Dion" chassis was conceived with great

---

*AIACR stands for the Association Internationale des Automobiles Clubs Reconnus. This group preceded the FIA (Fédération Internationale de l'Automobile).

versatility in mind. It was adaptable to rear-wheel drive or front-wheel drive and would accept engines from four bangers to narrow-angle V-16s. It was at home on dirt or pavement. But what probably thrilled Miller beyond imagining was that the design was made in heaven for the delivery of power to all four wheels. As we have seen, he knew that the historic moment for that was at hand.

Origins of Downdraft Carburetion

The arrival of downdraft carburetion on the American auto racing scene coincides with the introduction of the Winfield Type S carburetor very early in 1930. It was quickly followed by the SR, which incorporated an acceleration pump. Easily detachable float bowls permitted the throttle body and bowl to be mounted in updraft or downdraft position. Unlike Miller carburetors, these were designed to provide both top performance and top fuel economy. They quickly took over and, with the limited fuel formula that went into force in 1933, their use became virtually universal.

Edward A. Winfield was one of the greatest talents in the history of American motor racing. He was born in La Canada, California, in 1901 and went to work for Miller in 1915, at the tender age of fourteen. He began in the carburetor factory and then worked in the engine section until 1921, when he resigned in order to set up on his own account. While earning a reputation as a leading dirt-track driver on weekends, he became the national camshaft wizard before adding carburetor manufacture to his diversified activities. He died a legend in 1982.

In June 1994 I had the privilege of going through a portion of the personal photo collection of Col. E.J. Hall. The collection is preserved by Jay Eitel, nephew of the chief engineer of the Hall Scott Motor Co. Having been an eyewitness to the landmark arrival of downdraft carbs in 1930, it was with real shock that I came upon a beautiful photo of a neat marine installation of a Miller Golden Sub-type engine of about 1918. It sported a highly sanitary inlet manifold fitted with dual downdraft Miller carburetors. Like the Winfields of a dozen years into the future, the float bowls of the standard Miller were detachable and the throttle body was invertible—evidence of an idea that Harry Miller tried and tossed away.

Chapter 4

Born in the Crucible of Speed

Between the two World Wars, racing cars in the U.S.A. were almost always owned by small private enthusiasts and entrepreneurs. The Four Wheel Drive Auto Company was an exception, in that it was a good-sized industrial firm, with an appropriate sense of purpose and responsibility to its shareholders. The company regarded its very own racing car as an asset like any other, and the car warranted a drawer of its own in the company's active files. All correspondence, invoices, memos, and pertinent scraps of paper were filed therein, year after year, in a normally businesslike manner. This went on from 1931 through 1952, when the loan of the car to Bill Milliken (Chapter 18) expired and the old machine became a museum piece in Clintonville. Thus accumulated an archive of ultimately authentic documents that is unique in the world and is of dazzling historical merit.

To its profound credit, the Four Wheel Drive Auto Company preserved this mass of obsolete paper, down to and including check vouchers and old business cards. Following the firm's absorption by the Seagrave Fire Apparatus Co., Buck Boudeman persuaded its management to let him take over the preservation of the old car, and this unique trove of documentation went with it. And when, in 1986, Bob Rubin convinced Boudeman that he was the person to restore the car to its greatest glory, the archive was included. It is thanks to these phenomenally fortunate acts of conservation that we are privileged to reconstruct the historical record in the following manner:

On July 6, 1931 Olen acknowledged receipt of Harry Miller's "Dear Friend" letter of the 3rd. The mail moved in those days. A lawyer before he became a captain of industry, Olen rarely used two words when one would do.

July 6, 1931

Mr. Harry Miller
2843 E. Grand Blvd.,
Detroit, Michigan.

Dear Sir:-

Your letter of July 3rd arrived today with photographs of the car and your motor.

This matter will come up before our Board of Directors at their meeting on Tuesday, July 14th, and you will hear from me shortly thereafter.

Yours very truly,
FOUR WHEEL DRIVE AUTO CO.
President.

W.A.Olen,
EL

On the 15th he replied to Miller's rambling three-page scenario for the future in two crisp sentences; however, he did thaw in his parting salutation. The message spoke of "our" building such a car, not Miller. It specified a racing car for advertising, which rid the discussion of any thought of luxury supercars or of LSR (Land Speed Record) machines. It said, "Don't call us; we'll call you."

July 15, 1931.

Mr. Harry Miller,
2843 E. Grand Blvd.,
Detroit, Michigan.

My dear Mr. Miller:-

Replying to your letter of July 3rd, the Board of Directors authorized our building a Four Wheel Drive racing car for advertising purposes. There are certain conditions in connection with the matter which I will clear up and then you will hear from us again.

Yours very truly,
FOUR WHEEL DRIVE AUTO CO.
President.

W.A.Olen,
EL

The car and engine that Miller had proposed would be novelties to the world. The race was ten and a half months away.

Olen sent a company representative in Indianapolis to the Speedway office to find out what costs were likely to be involved.

Aug.6th,1931

The Four Wheel Drive Ayto Co. [sic]
Clintonville, Wis.

Att. Mr. W.A,Olen

Dear Mr Olen;

As per your request I ween [sic] over to the speedway office and asked regarding the gost [sic] of entering and getting a car ready for the race. I am enclosing a last years entry blak [sic]. There will onle [sic] be miner [sic] changes in the requirments [sic] next year.

The entry fee is $100.00. Garage rent at the speedway is free. As to mechanics that is a hardon [sic] to answer. They say that you could get some real good men for $30.00 per week plus a percentage of the winnings if you have a good car.

They feel that $1,000.00 will cover all expenses including gas oil tires salary and entry fee.

From

H Engel

The men at Gramercy Place had perhaps forgotten that the Four Wheel Drive Auto Company ever existed when, out of the blue, came a terse order for "one (1) Four-wheel-drive racing car."

October 26, 1931.

Mr. Harry A. Miller,
6233 Gramercy Place,
Los Angeles, California.

My dear Mr. Miller:-

Replying to your letter of July 3rd, 1931 and pursuant to a resolution of our Board of Directors, you are hereby authorized to build for us one (1) Four-wheel-drive racing car with an eight-cylinder engine to incorporate our type of front axle universal joints and our type of center differential.

The car, complete in every respect, is not to exceed $15,000.00, and is to be completed so that after all necessary adjustments, it will be ready for the race at Indianapolis on May 30th, 1932.

Yours very truly,
FOUR WHEEL DRIVE AUTO CO.
President.

W.A.Olen,
EL

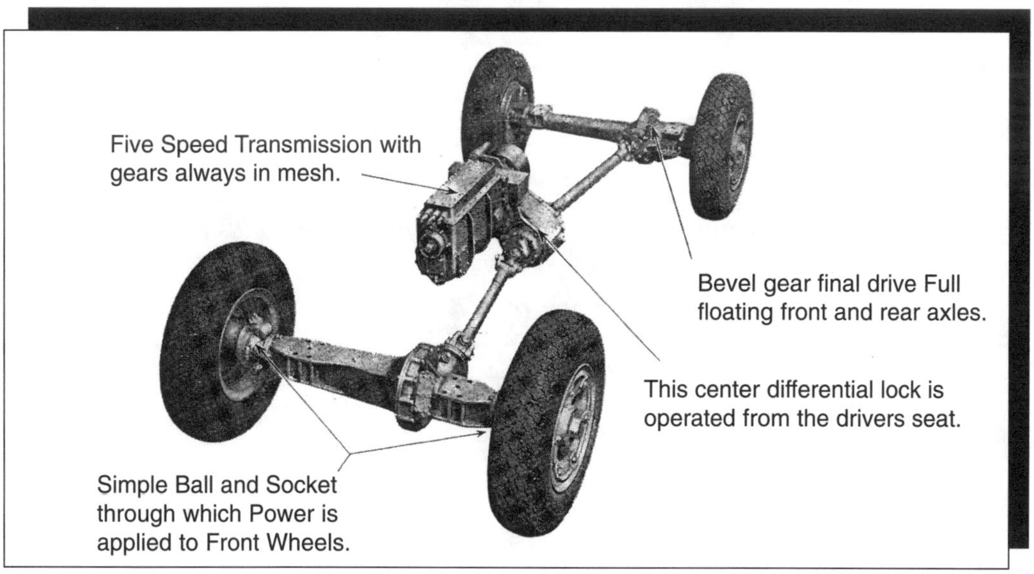

Illustration 4-1
Adverstisement for drive design Olen authorized Miller to build.

This legal instrument* naively required that the car "be ready for the race on May 30th," which is a rather late completion date. It specified "our type of universal joints" for the front axle, which probably would be mortal in racing; and, "our type of center differential," which it did not get, but which might have changed the course of history. A note on the file copy reads, "Pd $1000.00 10.26.31." The race was seven months away.

The next document in the files is a Western Union "night message" [telegram] from Miller to Olen, which reached Clintonville on the morning of December 1, 1931. The paltry $1000 had already vanished into salaries and materials and the monthly installment on the purchase price had not arrived. Six months to go.

11-30-31

Four Wheel Drive Auto Co.,
Clintonville, Wisc.

Attn; Mr. W. A. Olen

Am pleased to report that we are progressing very nicely on your order. Very anxious to get car completed soon as possible. Would appreciate your advancing Two thousand dollars to apply on contract.

Harry A. Miller

Chg.acct. Harry A. Miller, Inc.

More than two weeks had gone by and there had been no response to the telegram. Miller's secretary, T. A. Heinold, penned a progress report and a very polite reminder to the client. He was allowed to know that the supercar project was getting exciting.

December 16th, 1931

Four Wheel Drive Auto Co.
Clintonville, Wis.

Gentlemen: Attn: Mr. Walter A. Olen.

Enclosed please find copy of telegram sent to you under date of November 30 to which we have not yet received a reply.

We are pleased to report that we are progressing exceptionally well with the work on your car. All the detailed drawings are completed as well as many of the patterns. Many of the forgings have also been ordered. In view of this progress, we would appreciate having you send us as soon as possible your check for $2000.00 to apply on account.

*In the entire archive, there is no reference to a formal contract.

We are making every effort to complete this car as quickly as possible so that we will have ample time to make any changes if necessary. It is our intention to test this car on the Muroc Dry Lake and Mr. Miller will drive it himself.

Due to the small margrin of profit on special work of this nature and the general business depression, we will be somewhat handicapped to borrow money to carry on your work, therefore your kind consideration of mailing us your check for the above amount will be appreciated.

With the progress of this car, we are more enthusiastic than ever over its possibilities, particularly so for the pleasure car .

Trusting that you will give this matter your immediate attention, we are

Yours very truly,
HARRY A. MILLER, Inc.

T. A. Heinold
Secretary
TAH:TM

Surface mail moved between Los Angeles and Clintonville in two to three days. It took the client twenty-six days to react to the telegram. Everyone was in the same Depression boat.

December 26, 1931

Mr. Harry A. Miller, Inc.,
6233 Gramercy Place,
Los Angeles, Calif.

Dear Sir:-

Replying to your letter of December 16th, we are enclosing check for $2000.00. This makes a total payment on your account of $3000.00.

We are hoping, that you will be sending us some of the drawings to show just what the construction is.

Very truly yours,
FOUR WHEEL DRIVE AUTO CO.

W.A.Olen, President
MCK
Encl.

After a month of silence Miller wrote to Olen on January 26, 1932.

January 26th, 1932

Mr. W. A. Olen,
Four Wheel Drive Auto Co.,
Clintonville, Wisc.

Dear Mr. Olen;

We are mailing under separate cover a blue print showing how your car will look when completed.

We are progressing very nicely with the work on this car and at a later date will send you photographs showing different parts.

Assuring you that we will keep you advised regarding the progress on your job, and with kindest personal regards, I remain,

Cordially yours,

Harry A. Miller

HAM:TAH

(handwritten post script)
Blue prints in Eng. Dept. c/o Mr. Dodge.

Then, the better part of another month went by and Miller's secretary addressed to Olen another plea for funds.

February 24th, 1932

VIA AIR MAIL

Mr. W. A. Olen
Four Wheel Drive Auto Co.
Clintonville, Wis.

Dear Mr. Olen:

We are pleased to report that we are making very good progress on the Four Wheel Drive racing car which is on order for your company. Many of the major parts such as front axle, rear axle, transmission, frame and cross members, etc. are well under construction. Some of the assembly work is also in progress.

During your convesation with Mr. Miller regarding this car, we do not know whether or not he commented on our usual terms of payment covering such orders as this, Due to the nature of this special work, it has been our custom to require the various payments to be made in monthly installments.

Under ordinary conditions, we could no doubt make the necessary arrangements to carry on this work, but as this is very difficult, we would like to know if it would be possible to arrange to send us an additional payment of $5,000.00 to apply on account.

We would also like to know what color you have in mind to paint this car, and also what color upholstering you prefer.

Trusting that you will find it convenient to send us the above referred to check, and anticipating your early reply, we are

Very truly yours,

HARRY A. MILLER, Inc.

T.A. Heinald

T.A. Heinald [sic]
TAH:TM

The lack of communication between the parties during this period is remarkable. Receipt of this message was acknowledged on the 27th; Olen was away, and would reply upon his return. The race was about three months away.

February 27, 1932.

Harry A. Miller, Inc.,
6233 Gramercy Place,
Los Angeles, Calif.

Attention -Mr. T. A. Heinold

Gentlemen:-

Replying to your letter of February 24th, Mr.Olen is out of the city and a copy has been forwarded to him. We expect that he will return in about a week and the matter will have his attention.

Yours very truly,
FOUR WHEEL DRIVE AUTO CO.

ELV

Then came a fascinating letter to Olen from Ira Vail, ex-race driver become promoter. He had the four-wheel-drive idea years ago, he said, and he would like to be team manager for the new car. In addition to his three-page letter, he included another three pages of personal character references. The signatories were a veritable "Who's Who In American Championship Auto Racing."

February 29, 1932

Mr. Walter A. Olen, President,
The Four Wheel Drive Auto Company,
Clintonville, Wisconsin

Dear Mr. Olen:

Complying with your suggestion made in Boston when I called to see you with reference to my managing a racing car Harry Miller is constructing for your company. I will briefly outline a proposition for your consideration. I am also enclosing a list of references.

For several years I have had the idea of building a four wheel drive racing car, feeling that it would, without a question of doubt, be the fastest car ever built for track racing. I based my idea on the fact that the present type of race cars, front or rear drive, spin their wheels when accelerating, especially on a dirt track. The front drive cars lose so much traction that they are useless on a dirt track. The four wheel drive would be ideal for this work, as there would be no slippage, all four wheels taking the power evenly, eliminating the tendency of a car turning sideways when accelerated, as a rear drive does.

Last December I wrote Mr. Miller regarding the building of a four wheel drive car, outlining my views and asking for his opinion of the same. He replied that he was already building two four wheel drive cars, but did not say who they were for. As I explained to you in Boston, I called on your Mr. Daniels, as you suggested to me in a letter some time ago. I wanted to get his views on this type of drive and he informed me that you had placed an order with Miller for one of the cars now under construction. The thought then suggested itself that you would need a man of my experiences to manage and campaign your car so that the best results might be obtained.

For over twenty years I have been engaged in the racing business and its various branches. During that time I built and have had built for me more than a dozen racing cars. I have driven in about a thousand races and for many years have had the title of dirt track champion. Many of my records still stand today, although I have not been driving for several years. During 1916 and '17 I drove rebuilt stock cars for the Hudson Motor Car Company with marked success. I bought the first 8-cylinder car designed by Harry Miller, which was the starting of his success as a race car builder.

Five years ago my racing manager died. At this time I decided to carry on his work promoting races and giving up driving. I was appointed Director of the automobile races for the N.Y. State Fair at Syracuse. This is the second largest race in the U.S., only exceeded by the five hundred mile race at Indianapolis. I also manage races for six or eight of the better fairs in the East during September and October.

Before entering your car in any races I would make sure of its sound construction by making private track tests in California, after which it could be entered at Indianapolis for their May 30th race and the following championship races: Detroit June 5th; Altoona June 12th; Chicago June 19th; Syracuse July 4th and Boston (Rockingham Speedway) August 7th. Following these events it could be raced at the eastern fairs. I would be in an excellent position to feature this car and get it a huge amount of publicity and advertising. This would also include publicity for the truck as the originator of the four wheel drive principle. Both the car and truck could be exhibited on the Fair Grounds during the week at no extra cost to the company. This I am sure would attract a great deal of attention, as these fairs are visited by crowds ranging from one to two hundred thousand people.

In order to carry out a successful season, it would be necessary to get the car completed as soon as possible. Lack of sufficient time to properly prepare a car for racing is usually the cause of many failures, so I urgently suggest that you investigate my references and make a decision on the advisability of using my services. I believe that the mechanic that is going to keep the car in condition during the racing season should also work on the car during its construction, or at least when the car is finally assembled, so that he will be thoroughly familar with its every part.

The terms for my managing the car and its crew could be arranged on a percentage of the money won with the car, plus a small salary and expenses, or on any other basis suitable to you.

With the confidence that I have in four wheel drive plus my experience, I would feel certain of the greatest success possible.

Awaiting your early reply, I am
Most sincerely,

Ira Vail

Ira Vail
IV/C

Olen was impressed, and addressed this stranger as "My dear."

March 3, 1932.

Mr. Ira Vail,
80 Bellows Lane,
Manhasset, New York.

My dear Mr. Vail:

I have your letter of February 29th with reference. I have a Board of Directors' meeting next tuesday. We may come to a decision with reference to you at that time, however, there is some probability of my going West. If so I shall probably wait until I get out on the coast before making a final decision.

I am interested in your proposition and you will hear from us at a later date.

Yours very truly,
FOUR WHEEL DRIVE AUTO CO.

W.A. Olen,
EL

President

The question of car livery was answered by the Four Wheel Drive Auto Company's chief engineer.

March 4, 1932

Mr. Olen
General Manager:

Regarding the painting of car being built by Mr. Miller.

We have recommended that the main color be a light chrome yellow. The up-holstering to be dark brown and the striping to match the up-holstering.

Yours very truly,

H.B. Dodge,
Chief Engineer.
LAN
CC: Mr. Bard
CC: Mr. Hansen

Olen replied to the Miller letter of February 24. He was the first person to admit that he and his organization knew nothing about racing, but he had learned something from the articulate and all-business Vail. He asked for comment on Vail and now wanted the car to be running by April 15. That was less than six weeks away, and all that he had to show for his money was a blueprint.

March 4, 1932.

Harry A. Miller, Inc.,
6233 Gramercy Place,
Los Angeles,Calif.

Gentlemen:

I have just returned from a trip through the East and find your letter of February 24th.

The understanding with Mr. Miller at the time the contracts were made, was that we would make payments as the work on the car progressed. Reports were to be made to us at least monthly showing just what progress had been made. It was very definitely agreed that the car would be ready so it could be thoroughly tested out in the West before being brought to Indianapolis for the race. I will call your attention to Mr. Miller's letter of July 3rd and our letter of October 26th.

We are buying this car for a very definite purpose and for a very definite time.

I am enclosing check for $3000.00 to apply on account, - will remit more when you show progress justifying further payments.

We would suggest that the color be a light chrome yellow; the upholstering to be dark brown and the striping, to match the upholstering.

I expect to be in the West in the near future. The car should be in running order by not later than April 15th.

I have had several letters from a Mr. Ira Vail,80 Bellows Lane, Manhasset,New York. He desires to take charge of the entering of our car at different races. I have given him no information or assurance that we would be interested. From my talk with Mr.Miller I thought he would be able to make all these arrangements. I will be glad to have any comment you may care to make on Mr. Vail and the advisability of having him represent us.

The mechanic that is to take care of the car after it is completed probably should be selected now in order that he might become thoroughly conversant with all the parts of the car. I presume that additional parts are being made up so that they can be used in case of emergency.

I would like a statement from you when, in your opinion, this car will be ready for its first test.

Yours very truly,
FOUR WHEEL DRIVE AUTO CO.
President.

W.A.Olen,
EL

Encl.

Miller replied to Olen with the usual reassuring words concerning progress. He ignored the subject of Vail and his hijack play. He let it be known that he was in charge of campaigning the car until the Indianapolis race, "after which other arrangements can be made." This was maddeningly vague. He ignored Olen's very crucial closing request. Relations were going from bad to worse.

March 14, 1932

Mr. W. A. Olen
Four Wheel Drive Auto Co.,
Clintonville, Wis.

Dear Mr. Olen:

This will acknowledge receipt of your letter and check amounting to $3,000.00 for which kindly accept my thanks.

I am glad to state that we are proceeding exceptionally well with your car. While many of the parts are already completed, the general assembly work has been held up pending the completion of miscellaneous other parts.

I have selected a driver for your car as well as a thoroughly competent mechanic who has been working for me for some time.

In accordance with our conversation I am making all the necessary arrangements for the Indianapolis race after which other arrangements can be made.

Assuring you that we are working diligently to complete this car and anticipating your visit to the coast, I am,

Yours, very truly,

Harry A. Miller

Harry A. Miller

While omitting any reference to Vail in his letter to Olen, Miller wrote to Vail, telling him who was in charge. Vail's interpretation of the car's status after Indy was precise, whereas Miller was perfectly ambiguous.

The U.S. mail also moved with lightning speed between Manhasset, N.Y. and Clintonville. Olen replied by return mail to Vail, whom he realized might be able to fill a disturbing void in his life. He wasted words, talking around, and not about, what was worrying him.

March 25, 1932.

Mr. Ira Vail
80 Bellows Lane,
Manhasset, New York.

Dear Sir:-

I returned last night from a trip through the south and have your letter of March 23rd. I will be home until probably next Tuesday or Wednesday when I expect to start on a trip through the west. I am going through the northern tier of states and then down to San Francisco and Los Angeles. It will probably be between two and three weeks before I will be in Los Angeles.

I have a letter from Harry Miller in which he outlines the situation as I understand he did to you, however, I will try to get all matters cleared up and let you know as soon as I arrive in Los Angeles. I trust you will be patient in the matter as there are many angles to this that I have to iron out.

Yours very truly,
FOUR WHEEL DRIVE AUTO CO.

W.A.Olen, President

EL

On April 19, the Four Wheel Drive Auto Company mailed a $3000 check on account to Miller, with a one-sentence cover note.

April 19, 1932.

Harry A. Miller, Inc.,
6233 Gramercy Place,
Los Angeles, Calif.

Gentlemen:-

Attached hereto find check for $3,000.00, to apply on account.

Yours very truly
FOUR WHEEL DRIVE AUTO CO.

Secretary.

FG*HW
Encl.

Nothing was said about the April 15 deadline for the car being operational. Olen had been to the Miller plant and knew where things stood. Immediately thereafter, Harry Miller wrote to him concerning entry formalities and the naming of the car.

April 21, 1932

Mr. Walter A. Olen, President,
Four Wheel Drive Auto Co.,
Clintonvllle, Wisconsin.

Dear Or. Olen:

Inclosed [sic] please find a copy of the entry blank covering the four wheel drive race car for the Indianapolis race.

You will note that the name of the car is being left blank. It will be necessary that you send both the name of the car and also the check covering the entry fee to the Indianapolis Motor Speedway Corporation same to arrive not later than May 2nd, 1932.

I overlooked discussing with you while you were here the matter regarding the naming of the car but wish to assure you that it can be referred to by using any name you desire, for example:

Miller Special
Miller Four Wheel Drive Special
Four Wheel Drive Special

It being optional but be sure to forward this information to Indianapolis prior to May 2nd.

I have forwarded the original entry blank to the Indianapolis Motor Speedway Corporation and have advised them that you would send your check and the name of the car to them direct.

The inclosed [sic] entry blank is a copy only and for your files.

Yours very truly,

Harry A. Miller

Harry A. Miller

A second letter from Miller to Olen dated April 21 covered a very different range of subjects. It contained the first reference in this archive to the duplicate four-wheel-drive racing car. Leo Goossen told me that Miller, in electing to build this machine, gambled on his conviction that he would be able to sell it, too, to the Four Wheel Drive Auto Company. The 16-cylinder racing car was the one that was driven by Shorty Cantlon at the Speedway in 1931, and entered in the name of Bill White. The signs of financial strain were becoming dire.

OFFICIAL ENTRY BLANK

INDIANAPOLIS MOTOR SPEEDWAY CORP.
INDIANAPOLIS, INDIANA

Enclosed is remittance of $200.00 entry fee for one car in race meet described herein.

Date APRIL-18TH 1932

Name of Car ______ Model and Year V-8-1932

No. of Cylinders 8 Bore in Inches 3.500 Stroke 4.00 Inches

Piston Displacement 307.87 Cu. In. Supercharger? No Weight ______

Front ~~or~~ and Rear Drive FOUR WHEEL DRIVE Four-Cycle or Two-Cycle Principle? FOUR CYCLE

Gasoline—Diesel—Semi-Diesel? GASOLINE A.A.A. Registered Car No. ______

Name of Owner ______

Address of Owner ______

Name of Driver ______ Register No. ______

Address of Driver ______

Name of Mechanician ______ Register No. ______

Address of Mechanician ______

It is hereby understood that the car described above and driver named are to appear at the above-mentioned meeting, except as provided in the rules of the Contest Board, and particularly Rules 99 and 217 thereof.

All rules of the Contest Board, A.A.A., and the approved supplementary regulations are hereby made a part of this contract.

Signature of Entrant Four Wheel Drive Auto Co

P. O. Address Clintonville-Wis

Entries Close Midnight, May 2, 1932

Illustration 4-2 Entry blank for the four-wheel-drive car with name of car left blank.

Apr. 21, 1932

Mr. Walter A. Olen
Four Wheel Drive Auto Co.
Clintonville, Wisconsin

Dear Mr. Olen:

This will confirm our verbal understanding of last year wherein I have agreed to assume the responsibility of the management of your race car up to and including the Indianapolis race of 1932. It will not be necessary for you to have any managers, etc., until after this race.

As you have already been informed the other four wheel drive car which is a duplicate of the one we are building for you, is being offered at a price of $15,000.00 and I would like to see your company render a favorable decision in this matter because I believe it will be to your advantage.

If this cannot be arranged would it be possible for your company to advance in the form of a loan $10,000.00 and as security I could offer both the four wheel drive race car and the 16 cylinder race car. This would be a short term loan.

I also understand from Mr. White that during your conversation with him nothing definite was arrived at and in this connection would suggest that this matter be held in abeyance until after the Indianapolis race at which time White, Oldfield, Vail and others will be present.

In addition to the $3,000.00 which you had agreed to wire your factory for while you were in Los Angeles I would appreciate having you arrange to send another payment of $2,000.00 about a week later as it will help me materially in completing this car.

Thanking you for your kind consideration in this matter and awaiting your early reply, I am

Yours very truly,

Harry A. Miller

HARRY A. MILLER (Signed)

E.M. McLean of the Four Wheel Drive Auto Company service agency in Los Angeles reported to Olen on April 22 that much progress had been made on the car and that Miller "said that they would have it practically completely assembled by a week from today."

A letter dated the following day to Olen from T.E. "Pop" Myers, vice president and manager of the Speedway, acknowledged receipt of the Four Wheel Drive Auto Company entry filed by Harry Miller. Myers added, "We desire to assure you that we appreciate this entry very much and have every reason to believe that it is going to create a wide interest, not only among lovers of automobile racing but the members of the engineering fraternity as well." This, of course, was one hundred percent fact, as well as being music to the ears of the entrant.

Illustration 4-3 Harry Miller and Barney Oldfield with the Four Wheel Drive Company car in front of the Miller plant. (GBC ex FWDCO)

Illustration 4-4 The duplicate car mentioned in the Miller letter of April 21, 1932. (RMRC ex FWDCO)

On April 26, McLean wrote to Olen concerning a meeting that he had just had with Miller. Miller hoped to have the car ready to head east in eleven days. At this late date he had the good idea of transporting the racer on a Four Wheel Drive Auto Company (FWD) truck. He also introduced the novel idea of having retired veteran Barney Oldfield act as nominal team manager. McLean was not aware of the potential power of this proposed stroke of showmanship. Miller was living in a tomorrow in which a four-wheel-drive LSR car, driven by Oldfield, would help launch the production in series of his four-wheel-drive super road car.

April 26, 1932

Mr. W.A.Olen, President
Four Wheel Drive Auto Co.
Clintonville, Wisconsin

Dear Mr. Olen;

In response to Mr. Millers request I went out to his place today, he thinks it would be a good idea to have Barney Oldfield act as manager of the racing car at Indianapolis on a basis of 25 % of the winnings.

Miller contends that Oldfield as manager would materially increase the publicity as he is going east soon and would be talking the FWD idea where ever he went. The car Miller is figuring on to beat the present worlds speed record will be a four wheel drive and as he told you when you were here, Barney Oldfield will probably drive it, in fact he is going to build it for Oldfield and one of the reasons Oldfield is now going east is to raise the money to have the car built.

Miller states that even though Oldfield acts as manager at Indianapolis, he, Miller will select the driver and handle the arrangments as he agreed to do when you were here. Oldfield, as far as I can see, would only be valuable to the extent of what ever additional publicity his being manager would bring.

I understand from Miller that Oldfield is at present broke, my personal opinion is that Miller being a friend of Oldfield is anxious to see him make a little money, I think his bringing up this question of Oldfield acting as manager at Indianapolis is by way of doing Oldfield a favor rather than any additional advantage you would get.

Miller stated that whether or not you decide to have Oldfield, he will attend to all the details as per his agreement with you.

Illustration 4-5 FWD truck fitted to carry car to Indianapolis. (RMRC ex FWDCO)

Miller also brought up the question of transportation of the car to Indianapolis, he advises me that these racing cars are usually transported on truckz, [sic] they have a Ford truck with a flat rack body ready now for carrying the 16 cylinder car you saw on the floor, Miller thought it would be a good idea to use an FWD to transport the FWD racing car, he states this could be fitted with appropiate signs and would be a good advertisingng medimum [sic] going across the country. If you wanted to do this we could use the HH-6 we have in stock.

A body could be made and mounted for not to exceed $100.00.

Upon reciept of this letter please wire me what you desire to do in respect to the managment of the car at Indianapolis also whether or not you would care to use the FW, HH-6 we have in stock to transport the car across country.

They are assembling the car now and Miller stated he thought he could get it started for the east by May 8th. It is going to be painted yellow as you instructed.

Upon reciept of a wire from you I will get in touch with Miller again and advise him as to your decision.

Very truly yours

EM McLean

EM McLean

On April 26, Olen sent Myers the $200 entry fee for the car and informed him that it was to be known as the "Miller FWD Special." Still, on April 26, Olen sent "My dear Mr. Miller" his Board's denial of a $10,000 advance for the completion of the car. In this letter, certain elements of team management were defined.

April 26, 1932.

Harry A. Miller, Inc.,
6233 Gramery Place, [sic]
Los Angeles, Calif.

My dear Mr. Miller:-

Replying to your letter of April 21st, the car will be named "Miller FWD Special" and we are so informing the Indianapolis Motor Speedway Corporation.

We hereby confirm the understanding as expressed in your letter that you will look after the management of the car until after the Indianapolis race 1932.

We authorize you to allow the mechanic $50.00 per week during the race, and, in addition, 10% of any winnings. Also to employ a driver and to allow him 33-1/3% of the winnings, and a driver mechanic at 5% of any winnings. It is our understanding that, no other compensation or expense is to be taken care of by us.

Illustration 4-6 Barney Oldfield, named as team manager, perched on Miller's Golden Submarine about fifteen years earlier. (FAUC)

The freight on the car to Indianapolis, of course, will be paid by us.

It is our understanding that you will have contracts made with each of these several persons whom you select which will be so drawn that they will assume the responsibility for any personal injury to themselves that may be incurred by reason of the race.

Yours very truly,
FOUR WHEEL DRIVE AUTO CO.

W.A.Olen, President
EL

On the same date, in another letter, Olen amiably asked Miller to do him a few favors.

April 26, 1932.

Harry A. Miller, Inc.,
6233 Gramercy Place,
Los Angeles, Calif.

My dear Mr. Miller:

With further reference to the Indianapolis race and the "Miller FWD Special", we would appreciate your preparing a statement for the press that you think is the proper way of presenting this car and its outstanding advantages for this type of service.

We would like this statement for our own advertising and publicity department.

We would also appreciate your advising us how you think the publicity should be handled. We would like to have everything emanating from or through you as much as possible.

Yours very truly,
FOUR WHEEL DRIVE AUTO CO.

W.A.Olen, President
EL

On April 29, with qualifying now only twenty-three days away, Harry Miller sent Olen "photograph and photostat which are self-explanatory." He said that he would prepare "the article," and that assembly of the car had begun. At last, on May 1, McLean visited the plant, finding that the car might be ready to try out "some time this week." On May 3, Harry Miller wired Olen. Qualifying would start in nineteen days.

TELEGRAM

GUB 28 DL
LOSANGELES CALIF MAY 3 1932 1040A

WALTER A OLEN

PROGRESSING VERY FAVORABLY WITH CAR MUST HAVE SOME MONEY PLEASE SEND YOUR LOCAL REPRESENTATIVE BALANCE DUE ON CAR SO HE CAN PAY US AS WORK PROGRESSES VERY URGENT

HARRY A MILLER INC
120P

Not having received McLean's letter of the 1st, Olen wrote to him on the 4th, defining policy.

May 4, 1932.

Mr. E. M. McLean,
Four Wheel DriveSales Co.,
411 Stanford Ave.,
Los Angeles, Calif.

My dear Mr. McLean:-

You have not yet reported on just what progress has been made in the racing car. We are anxious to know at frequent intervals just how that car is coming along.

You are authorized to furnish the 2-ton truck to haul the race car to Indianapolis. Under separate cover you will receive the parts to change the gear ratio so as to provide more speed. I would suggest that the truck be driven to Harry Miller's place so he can look it over. He may want some kind of body on it. Perhaps some of the men would want to ride through in the truck, all of which, of course, would be permissible. You are authorized to aid in putting the truck in shape to make it suitable for the trip.

The race car has been entered at Indianapolis as the "Miller FWD Special".

We would prefer to make all payments and give all instructions to Harry Miller from here. The reason is that we do not want a divided responsibility. There are many hazards in a race and the conditions under which Miller and his organization are undertaking to handle this race may mean that they get very little out of it in which case there might be a tendency to want to claim some different contract with you than was made with the home office.

Mr. Miller appeared before the Board of Directors, stated the conditions under which he would handle the whole problem, and I am simply carrying through this understanding and I wish to be very careful that no side issues are brought in to vary or change the setup.

I might say that the principle [sic] object of having a car in the race is to show the public that four wheel drive construction is capable of dependable and controlled speed.

Your attitude on all matters, therefore, should be that you are simply cooperating with the home office in carrying out Mr. Miller's understanding with the home office direct.

Yours very truly,
FOUR WHEEL DRIVE AUTO CO.

W.A.Olen, President
EL

Olen to Miller: Still no press release, no contract forms for driver and mechanics. Oldfield was accepted grudgingly.

AIR MAIL

May 4,1932.

Harry W. Miller, Inc., [sic]
6233 Gramercy Place,
Los Angeles, Calif.

My dear Mr. Miller:-

I am enclosing check for $2000.00 to apply on account. We would prefer to handle all the payments of money from here.

We are authorizing our service station at Los Angeles to provide a truck with which you can drive through to Indianapolis with the racing car.

I have not permitted our organization to make any statement with reference to the racing car whatever because we wanted the statement to come from you. In addition to any publicity that you can get by sending the statement out yourself or broadcasting over the radio, or any other method that you have, we will be glad to send the statement out as you prepare it and with your signature in addition to our own. We are, therefore, anxiously waiting this statement.

We would also like to have the contracts for the drivers and the mechanics.

We have a letter from Mr. McLean in which he stated you suggested that we allow Barney Oldfield 25% for managing the car at the Indianapolis race. None of us have ever met Mr.Oldfield; none of us knows anything about racing and we dislike very much to make a shift in the plans at this time. You appeared before our Board of Directors and it is their confidence in you that has resulted in the car being built and entered in the race, and we would much prefer to have the program go through just as it was outlined at the beginning. We will, however, authorize you to allow Barney Oldfield up to 15% of any winnings of the car if, thereby, he can be a help to you, but we shall look to you not only for the completion of the car, for its delivery at Indianapolis for the race, but also for the complete management of the car during the entire race.

I trust that this will be satisfactory.

Yours very truly,
FOUR WHEEL DRIVE AUTO Co.

W.A.Olen, President
EL Encl.

With qualifying sixteen days away, Olen demands a press release.

May 6, 1932

HARRY A MILLER
6233 GRAMERCY PLACE
LOS ANGELES CALIF

SEND IMMEDIATELY AIRMAIL STATEMENT FOR PUBLICATION

OLEN

With qualifying now thirteen days away, Harry Miller sent Olen the press release for which he had been languishing. He sent a photo of the driver he had chosen, and stated that the car should be shipped on the 16th. The duplicate four-wheel drive and the 16C remained unsold.

May 9, 1932.

AIR MAIL

Mr. Walter A. Olen,
Four Wheel Drive Auto Co.,
Clintonville, Wisconsin.

Dear Mr. Olen:

Inclosed [sic] please find photograph of Mr. Arvol Brummier, who has won several races here recently and is my first choice to pilot your entry.

I am also enclosing herewith an article for the press. All through this article I have given credit to Barney Oldfield as being manager of the car but this is for publicity sake only, as more publicity could be obtained that way than otherwise.

As soon as the photographs of loading the car are taken we will be pleased to send you the negatives and also a list of the various newspapers who will be glad to run the picture and your article. This list of newspapers, totaling approximately 200 has been furnished by Mr. Bill Pickens, who was Mr. Oldfield's publicity manager.

We are releasing an article to the local representative of the N.E.A. which is similar in substance to the inclosed [sic] article. Their contacts will reach some 800 newspapers and no doubt your clipping bureaus will send you copies. We hope that this will meet with your approval.

The car ls coming along in an excellent manner, the body and parts are being painted and the final assembly work is rapidly nearlng completion. It is expected that the car will be ready for shipplng on Monday, the 16th. We are contemplating runnlng it on the local Legion Ascot Speeday [sic] prior to shipping.

As we are now working day and night completlng these cars, we find it very hard to carry on without additional money and would therefore appreciate your consideration be [sic] sending us another payment to apply on your order.

Up to the present writing, we have not sold either of our cars.

Sincerely hoping that the enclosed article will meet with your approval and anticipating your immediate reply, we are,

Yours very truly,

Harry A. Miller

Harry A. Miller

The Final State of Panic Sets In

May 16,1932.

Harry W. Miller, Inc., [sic]
6233 Gramercy Place,
Los Angeles, Calif.

Dear Sir:-

Please refer to the last paragraph of Mr. Olen's letter to you of April 26, 1932, regarding the relieving of this company of responsibility for any personal injury that may be incurred to the several persons in charge of racing car.

Up to this date we have not received the contracts. As the management is desirous of having these contracts, we trust we will receive them by return mail, if they have not already been sent to us.

Very truly yours,
FOUR WHEEL DRIVE AUTO CO.

HH*HW

Accounting Department.

GU13 20 DL=INDIANAPOLIS IND MAY 16 212P 1932

FOUR WHEEL DRIVE AUTO COMPANY=

CLINTONVILLE WIS=

NO DRIVER FOR YOUR CAR MUST HAVE THIS INFORMATION NOT LATER THAN THURSDAY ACCOUNT PROGRAM GOING TO PRESS PLEASE WIRE=

INDIANAPOLIS MOTOR SPEEDWAY CORPORATION.

249P.

May 16, 1932

Indianapolis Motor Speedway Corp.
Indianapolis, Indiana

Gentlemen:-

We have your wire of May 16th requesting the name of driver for our entry "Miller FWD Special".

We have wired Harry A. Miller to wire you name of driver and if you have not received this information by Wednesday, May 18th, please wire us.

Trusting the necessary information will be furnished you, we are

Very truly yours,
FOUR WHEEL DRIVE AUTO CO.

Accounting Department

HRH:T

GU 49 DL COLLECT XU=Q LOSANGELES CALIF MAY 16 1932 932A

FOUR WHEEL DRIVE AUTO CO=

CLINTONVILLE WIS=

NO PROVISION MADE FOR INSURANCE ON CAR MILLER STATES NOT CUSTOMARY TO INSURE STOP SHALL I ARRANGE FOR SAME STOP MILLER REQUESTS US TO ADVANCE ONE HUNDRED FIFTY DOLLARS FOR EXPENSE DRIVING TRUCK TO INDIANAPOLIS PLEASE WIRE INSTRUCTIONS REGARDING INSURANCE ALSO ADVANCE EXPENSES STOP TRUCK WILL LEAVE MONDAY=

E M MCLEAN.

1158A.

GU11 15 COLLECT=Q LOSANGELES CALIF MAY 16 1932 958A

FOUR WHEEL DRIVE AUTO CO=

CLINTONVILLE WIS=

INSURANCE ON RACING CAR SIXTY TWO DOLLARS STOP ONE HUNDRED DOLLARS DEDUCTABLE WIRE INSTRUCTIONS=

E M MCLEAN.

104P.

Barney Oldfield Discusses New Engineering Developments: America's Greatest Driver Praises Four Wheel Drive Creations

by Barney Oldfield, Dean of American Speed Kings

INDIANAPOLIS MOTOR SPEEDWAY, INDIANAPOLIS, IND.

The entries of the Four Wheel Drive Auto Company of Clintonville, Wisconsin, and Harry Miller of Los Angeles, California are the only cars in this year's 500-mile race which embody principles new to the speed world. They are attracting the spotlight of interest of the automotive industry and leading engineers from all over the world will be on hand at the great Indianapolis proving ground to see these distinctly new speed creations perform on Decoration Day.

These two cars are fitted with a V-Type Eight-Cylinder motor, the latest product of Harry Miller's genius and both use the principle of power application to all four wheels which has been used so successfully by the Four Wheel Drive Auto Company in their trucks for the past twenty-one years. The two cars are identical in every respect.

The entry of the Four Wheel Drive Auto Company will be driven by Bob McDonogh, the brilliant young star who holds a dozen or more world's speed records including the average of 134 miles per hour for 250. McDonogh qualified the car easily at an average of better than 113 miles per hour for the four-time circuit of the two-and-one-half mile Indianapolis track.

Both McDonogh and Miller declare that the car is the fastest job that Miller has ever produced and are expecting to set a new record for the 500-mile classic.

This new creation brought to the Altar of Speed for its initial baptism in the grueling grind of the 500-mile race has attracted the particular interest of the automotive industry. Many leaders in the business believe that the four-wheel-drive principle of power application will be found in fine pleasure cars offered to the public within the next year or two.

The passenger car of tomorrow is the racing car of today. Four Wheel brakes, straight-eight cylinder motors, high compression heads, modern exhaust and intake manifold design, carburetion improvements and many standard features of today's pleasure cars were first developed in the speed creations entered in the 500-mile race.

Illustration 4-7 Press release.

DESCRIPTION OF MILLER FWD SPECIAL

The MILLER FWD SPECIAL has a wheelbase of 100″ and tread of 56″. Fully equipped it weighs 2325 lbs. The power plant is a V-eight cylinder with 3 1/2″ bore and 4″ stroke giving a piston displacement of 308 cubic inches and develops three hundred horsepower at 5500 revolutions per minute. The including angle of the V is 45 degrees, making an exceptionally narrow engine.

A five-bearing counter balancing crankshaft is used. Side-by-side tubular connecting rods, four overhead camshafts, two magnetos, and four downdraft carburetors attached directly to the top of the cylinder block are used. The weight of the engine is 540 pounds.

The drive to the front and rear wheels is accomplished by the means of propeller shafts from the transmission to the front and rear wheels. In the transmission there is a center differential which compensates for the difference in distance traveled between the front and rear wheels. It is obviously as essential at this location as the differentials in the axles. This differential is so arranged that it can be locked as desired, resulting in a positive drive from the transmission to the axles.

On the front axle the outboard joint is of the trunion and ring type, so placed that its axis is on the steering knuckle center. The steering knuckle pin is inclined so that a line extended through its center will coincide with the centerline of the tire at the ground.

The rear axle shafts have bronze block universals, both inboard and outboard. Two chassis springs are used per wheel, and are quarter-elliptic, mounted one above the other. There being no rigid member between the wheel or differential case, instead a dead axle connects the two wheels, both front and rear. The wheels become semi-independently sprung.

Among the advantages resulting from a four-wheel-drive racing car, passenger car or truck, are the following: No wheel spinning as maximum torque of engine is applied to all four wheels, better traction on oily or slippery tracks or roads, increased tire life as maximum stresses on tires are reduced, superior acceleration or get-away, turns or curves on tracks and roads can be taken at higher speeds with the greatest safety, due to entire elimination of rear-end whip. Bob McDonogh is very enthusiastic about the Miller FWD Special. He declared "a ten-year-old boy can handle this job, because it handles so easily and has no side slippage."

May 16, 1932

E M McLean

FOUR WHEEL DRIVE SALES CO

411 STANFORD AVE

LOSANGELES CALIF

HAVE MILLER WIRE INDIANAPOLIS MOTOR SPEEDWAY CORPORATION NAME OF DRIVER AS PROGRAM GOING TO PRESS THURSDAY ALSO ADVISE STOP SATISFACTORY ADVANCE ONE HUNDRED FIFTY DOLLARS AND ARRANGE FOR RECEIPTS COVERING EXPENDITURES DURING TRIP STOP TAKING CARE OF INSURANCE FROM HERE

FOUR WHEEL DRIVE AUTO CO

H HOARE

ELV

Miller's invoice for "platform body" for Four Wheel Drive Auto Company truck:

5/17/32
Four Wheel Drive Auto Co.
Clintonville,
Wisc.

1 - Special truck body built on Four Wheel Drive truck
Platform body to transport race car to Indianapolis
$96.19

[handwritten note]
Recd payment in full
May 27/1932

H.A. Miller

Miller statement of account, 17 May:

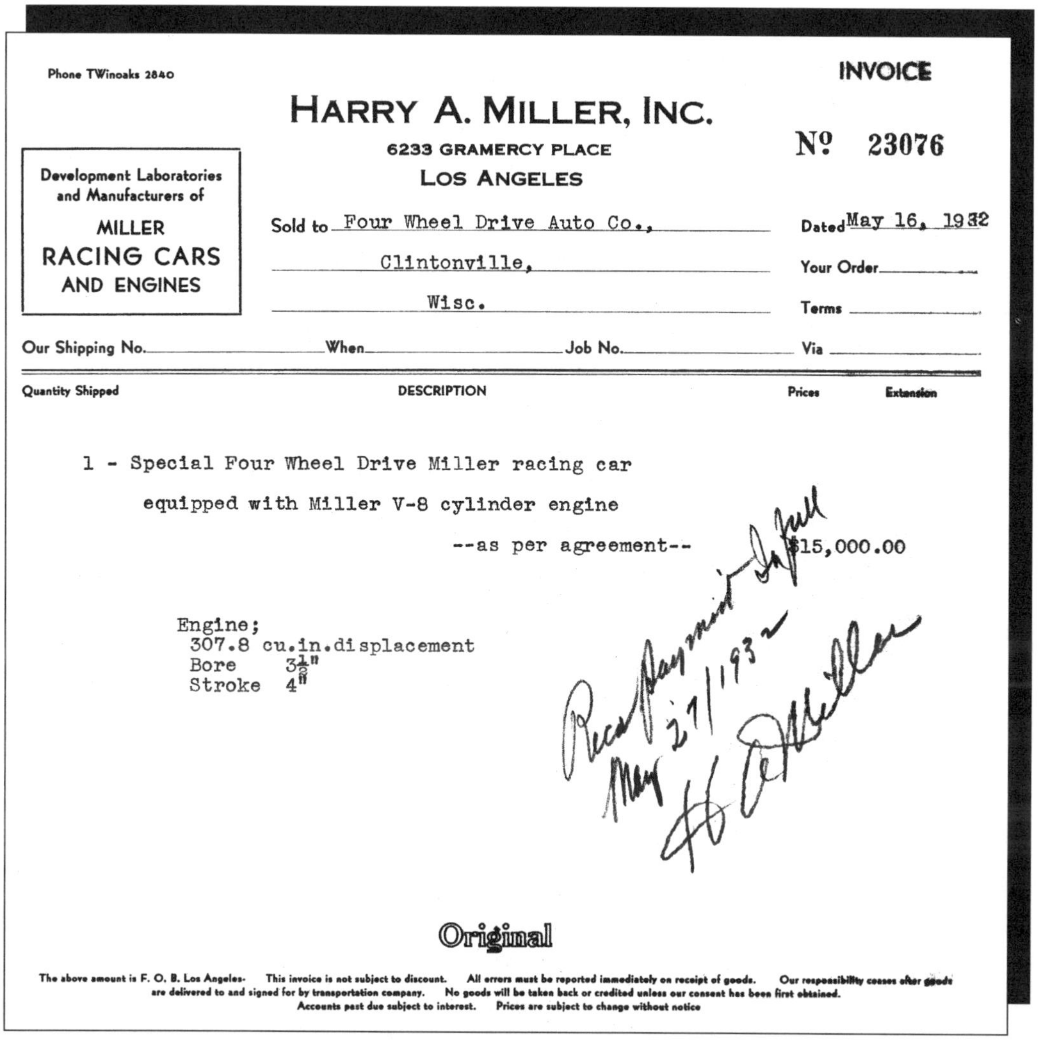

Phone TWinoaks 2840

INVOICE

HARRY A. MILLER, INC.

6233 GRAMERCY PLACE
LOS ANGELES

№ 23076

Development Laboratories and Manufacturers of
MILLER
RACING CARS
AND ENGINES

Sold to Four Wheel Drive Auto Co.,
Clintonville,
Wisc.

Dated May 16, 1932
Your Order
Terms

Our Shipping No. When Job No. Via

Quantity Shipped	DESCRIPTION	Prices	Extension
1 -	Special Four Wheel Drive Miller racing car equipped with Miller V-8 cylinder engine --as per agreement--		$15,000.00

Engine;
307.8 cu.in.displacement
Bore $3\frac{1}{2}$"
Stroke 4"

Rec'd Payment In Full
May 27/1932
H A Miller

Original

The above amount is F. O. B. Los Angeles. This invoice is not subject to discount. All errors must be reported immediately on receipt of goods. Our responsibility ceases after goods are delivered to and signed for by transportation company. No goods will be taken back or credited unless our consent has been first obtained. Accounts past due subject to interest. Prices are subject to change without notice

Illustration 4-8 Invoice for Special Four Wheel Drive Miller racing car.

STATEMENT

To: Four Wheel Drive Auto Co.,
Clintonville, Wisconsin.

5/17/32	Invoice	#23076	$15,000.00	
	"	#23077	96.19	
				$15,096.19
Credits				
Nov. 1, 1931	Cash		$1,000.00	
Dec. 31, 1931	"		2,000.00	
Mar. 9, 1932	"		3,000.00	
Apr. 23, 1932	"		3,000.00	
May 6, 1932	"		2,000.00	11,000.00
Net balance due				$4,096.19

Four Wheel Drive Auto Company statement with check numbers:

STATEMENT—
Harry A. Miller, Inc.,
6233 Gramercy Place
Los Angeles, Calif.

Invoice No.	23076	5/16/32- - - -	$15,000.00
" "	23077	5/17/32- - - -	96.19
			$15,096.19

Payments made:

Ck.	#41092	– 10/26/31 –	$1,000.00
"	42496	– 12/26/31 –	2,000.00
"	242	– 3/4/32 –	3,000.00
"	288	– 4/19/32 –	3,000.00
"	29	– 5/4/32 –	2,000.00
"	45289	– 5/19/32 –	1,200.00
			$12,200.00

Balance due - $ 2,896.19

(handwritten note) Pd 5/27/32

The truck with the race car left Los Angeles near midnight on the 16th.

GU17 36 DL COLLECT=Q LOSANGELES CALIF MAY 17 1932 1105A

FOUR WHEEL DRIVE AUTO CO=

CLINTONVILLE WIS=

NO DRIVER SELECTED AS YET MILLER WIRING INDIANAPOLIS ALSO YOU STOP TRUCK LEFT LAST NIGHT ELEVEN THIRTY WILL ARRIVE INDIANAPOLIS THIS WEEK STOP MILLER LEAVING TONIGHT BY TRAIN STOP I AM LEAVING TONIGHT FOR SALTLAKECITY=

E M MCLEAN.

129P.

Miller to Olen in Clintonville. Leaving tonight, money imperative.

GU4 37 NL LOS ANGELES CALIF MAY 17 1932

WALTER A OLEN

FOUR WHEEL DRIVE AUTOMOBILE CO CLINTONVILLE WIS

LEAVING TONIGHT ON TRAIN IMPERATIVE I HAVE TWELVE HUNDRED DOLLARS WOULD YOU PLEASE WIRE THIS AMOUNT TO T. A. HEINHOLD MY SECRETARY HERE STOP TO APPLY AGAINST BALANCE DUE STOP LOOKING FORWARD TO SEEING YOU AT INDIANAPOLIS

HARRY A MILLER

808A MAY 18

Four Wheel Drive Auto Company to Olen in Indianapolis: Still no driver.

May 17, 1932.

W A OLEN

CLAYPOOL HOTEL

INDIANAPOLIS INDIANA

WIRE FROM SPEEDWAY CORPORATION STATES NO NAME DRIVER RECEIVED PROGRAM GOING TO PRESS THURSDAY STOP WIRED MOLEAN HAVE MILLER WIRE SPEEDWAY CORPORATION STOP REPLY FROM MCLEAN NO DRIVER YET SELECTED MILLER WIRING INDIANAPOLIS ALSO US MILLERS WIRE NOT RECEIVED STOP CAR LEFT MONDAY NIGHT MILLER LEAVING TUESDAY NIGHT BY TRAIN

FOUR WHEEL DRIVE AUTO CO

H.HOARE
ELV

Miller to Olen in Clintonville. Oldfield as driver!

GU28 21 LOS ANGELES CALIF MAY 17 1932 302P

WALTER A OLEN

FOUR WHEEL DRIVE AUTO CO CLINTONVILLE WIS

SUGGEST ENTERING BARNEY OLDFIELD AS DRIVER YOUR CAR STOP DEFINITE DRIVER WILL BE SELECTED AFTER TRIALS STOP CAR LEFT LAST NIGHT

HARRY A MILLER

514P

Indianapolis Motor Speedway to Four Wheel Drive Auto Company: Miller did it. Barney Oldfield is the designated driver.

May 18, 1932.

Four Wheel Drive Auto Company,
Clintonville,Wis.

Gentlemen:
Attention: H.R. Hoare.

This morning we are in receipt of a wire from Harry Miller designating Barney Oldfield, as driver for the Miller Four Wheel Drive Special.

Very truly yours,
INDIANAPOLIS MOTOR SPEEDWAY CORPORATION.

T.E. Myers

TE Myers
Vice-President.
TEM:hr

McLean to R.H. Schmidt: He sent photos of car and truck, which went into the company file. We learn that the mechanic is Ernie Olson, one of the greatest ever. Should reach Speedway sometime on the 20th.

5-17-32

To R.H.Schmidt at Clintonville
From E.M.McLean at Los Angeles

Dear Rudy:

Am enclosing some films I took of the car and the truck, the pictures of the truck are the only ones I was able to get as it was not completed until 10 o'clock last night, Aleck and I spread the canvas over it late in the afternoon to get these pictures.

The truck left last night with two drivers who will rdive [sic] steadily, Aleck followed it about 40 miles out of town and reported it was making 45 miles and hour right along.

Olsen the mechanic is following the car in a Buick, he is accompanied by his wife who is taking turns driving.

They shouls [sic] be in Indianapolis Friday sometime.

We installed a speed ometer on the truck and instructed the drivers to keep an acurate record of gasoline consumption as I figure these fellows will establish a worlds speed record for four wheel drive trucks on this trip and the gas consumption will come in handy for future advertising purposes.

I figure they should arrive in Phoenix about four this afternoon.

I am leaving this evening for Salt Lake and do not know just how long I will be gone, will write you from there regarding conditions

Yours truly

EM McLean

EM McLean

Olen gave his instructions to his general sales manager, R.H. Schmidt, who was to be in charge of this operation.

Illustration 4-9 Ernie Olson played a key role in Jimmy Murphy's winning the French Grand Prix in 1921. He continued to be depended on by America's top drivers. (GBC ex Ernie Olson)

May 20, 1932.

Mr. R. H. Schmidt,
Sheffield Inn,
Indianapolis, Indiana

Dear Rudy:

I am informed that Mr. Brummier, who was originally selected to drive our car, is to be married or has recently married and his wife objects to his driving. This will necessitate selecting some other driver.

I think that it would be well for you to remain at Indianapolis until after the race and see that everything is lined up in proper shape.

I am enclosing copy of the original correspondence with Mr. Miller so you will have the story exactly as it is. I would not make any changes but it should be completed, that is, agreements with the drivers and others so that the company will be assuming no responsibility in case of injury. I do not care to have the car itself insured because it is ours and our own liability.

Mr. Meyer, who has charge of the Speedway, gave me the name of Ed. Gates of the firm of Gates & Gates as one competent to draw an agreement where the driver assumes responsibility. He also gave me the name of Chas. Roemler. In case that Miller has no regular form of contract, you would perhaps better have the attorney draw the agreement.

I would be on the alert at all times that we are not taken advantage of, as what I have seen of the Indianapolis Races in the past it is the attitude of everybody in Indianapolis to get all they can out of it.

I am suggesting to Mr. Hanson that he arrange with Mr. DeuPree to have his office so there will be a stenographer and telephone service and headquarters where people can meet. Mr. Hanson will have charge of the publicity end of matters and you will have charge of everything else. We will be communicating with you in care of DeuPree's office. If it is necessary to make a trip to some nearby town during the week, that, no doubt, can be done.

The car was held up in Texas for want of a license and we wired $73.00 to the driver to take care of it. That happened last night at about 5:00 o'clock, so I do not expect the car will be in Indianapolis until Saturday night or Sunday night. Miller asked for $1200.00 and it was wired to him yesterday.

Yours very truly,
FOUR WHEEL DRIVE AUTO CO.

President.
W.A.Olen,
EL

Mr. Miller would take care of the releases as a matter of course.

The evening of Monday, May 23, 1932, company advertising manager W.N. Hanson addressed nine pages of single-space typescript to Olen, accounting for what he had done since the previous Friday. With a factory employee named Bill Smith, he had made the trip from Clintonville to the Speedway in "Nancy Hank," a 1911 Four Wheel Drive Auto Company "passenger car" or flatbed truck. The vehicle was still in more or less current use, they claimed, and the publicity stunt served to let a generally uninformed public know that four-wheel drive was no newfangled gimmick. Hanson met famed Speedway publicity chief Steve Hannigan (whose name he persisted in spelling "Harrigan") and sports writer Bill Sturm. Sturm also was a powerful press agent and sports promotion figure, Frank Lockhart having been just one of his clients. The letter is devoted to the craft of the huckster, but contains some comic relief, as when Hanson wrote, "Harry Miller has never road in any of his racing carts on this or any other track." And, speaking of the Four Wheel Drive Auto Company, he quoted driver Bob McDonough (whose name he also consistently misspells)

Illustration 4-10 "Nancy Hank," a Four Wheel Drive Company vehicle of 1911, was driven from Clintonville to the speedway by employee Bill Smith, accompanied by advertising manager W.N. Hanson. (GBC ex IMS)

saying, "… it has entered my car with Barney Oldfield as team manager as a sporting jester …" (Author's Note: "Gesture," perhaps?)

There were pearls in the letter, such as, "I discussed publicity matters with Oldfield and he said that anything we did was agreeable with him and he offered to cooperate with us in any stunts or publicity schemes we could work out. Mr. Oldfield is a great publicity man himself in so far as working out ideas is concerned, but has not the temperment [sic] to direct or follow through on details, and of course is busy managing our entry."

Hanson summed matters up thus: "The outstanding thing of the race is the F.W.D. entry of its Four Wheel Drive racing car. We think we got the publicity ball rolling yesterday. There is much discussion regarding the F.W.D. possibilities throughout the racing fraternity. We have a wonderful set up. First we have the only really new thing in the race. We have Barney Oldfield still an outstanding figure in the racing world as manager, and a car built by Harry Miller whose ability is unquestionable and who has more cars in this race than any one else."

Also on May 23, Mr. Olen obtained the legal instrument that he had been seeking, and had it signed by the members of his racing team. (Editor's Note: Signatures are those of Olen, Frank Gause (secretary of company), and Barney Oldfield—hardly the racing team!)

THIS INDENTURE made and entered into the 23rd day of May, 1932, between the Four Wheel Drive Auto Company of Clintonville, Wis., hereinafter called "Lessor" and Barney Oldfield of Beverly Hills, California, Robert

McDonough of Detroit, Michigan, Herschel McKee of Indianapolis, Indiana, and Ernest Olsen of Los Angeles, California, hereinafter called "Lessees"

WITNESSETH:

That the Lessor hereby leases to Lessees, and Lessees hereby agree to rent from the Lessor, one racing car described and known as a "Miller Four Wheel Drive Special" now at the Indianapolis Speedway track for the term beginning May 23rd, 1932, and ending on May 30th, 1932, at 5 o'clock P.M.

It is expressly agreed and understood that said racing car may be driven by lessees, or any of them, in the 500 mile automobile race to be run at the Indianapolis Speedway on May 30, 1932, and that lessor will furnish all repair parts, oil, gasoline, tires, etc., required during the term of this lease, to put and keep said car in condition and to participate in said lease.

IN CONSIDERATION WHEREOF, the lessees bind themselves to pay as rent for same a sum equal to 36-2/3% of the gross prize winnings or earnings of said car earned as a result of the lessees participation in said race, said sum to be due and payable to lessor at the office of the Indianapolis Speedway immediately upon payment of said prize money or earnings.

The lessees further agree that said Miller Four Wheel Drive Special shall not be removed from the Indianapolis Speedway without the consent of lessor first had in writing and that at 5 o'clock P.M, on May 30, 1932, or as soon prior thereto as the 500 mile race is completed, or said car is declared out of said race, lessees will deliver said car to lessor in its then condition whatever it may be.

It is further agreed that lessor shall not be liable to said lessees for damages of any kind or from any cause whatever to the persons or property of lessees, or any person claiming by, through or under them, and lessees agree to defend at their own expense all damage suits which may be brought upon claims relating to said cause or the use thereof by the lessees.

IN WITNESS WHEREOF, the parties hereto have hereunto set their hands and seals the day and year first above written.

	(signed) Four Wheel Drive Auto Co., Lessor
Attest:	(signed) Walter A. Olen, President
(signed) Frank Gause, Secretary	(signed) Barney Oldfield, Lessee

A really neat stunt was the transport of the racing car on one of the company's current products, and under respectable conditions of performance, which were billed as "a record."

In another lengthy report to Olen, hand-written on May 24, Hanson explained that he had had new signs painted for "Nancy Hank" and the Four Wheel Drive Auto Company truck. "... in view of recent developments on the track we want it unmistakable who is entering the Miller-FWD Special." The "recent developments" were the arrival of the duplicate Miller four-wheel drive at the Speedway, and preparations for qualifying that car. With press agent Herman Dewpree, Hanson had written an article [full text in sidebar on pages 46 and 47], signed by Barney Oldfield, which was being sent to papers all over the country. Dewpree also was chief of the Speedway's press information staff and had arranged for "Nancy Hank" and the Four Wheel Drive Auto Company truck to be parked strategically behind the press stand on race day. The racing car qualified this day as follows:

Lap	Time	MPH
1	1:19.69	112.938
2	1:20.08	111.388
3	1:19.52	113.179
4	1:18.51	114.635
Time & average	5:17.80	113.279

Just before sundown the duplicate Miller four-wheel drive qualified at 112.003 mph.

Hanson reported to Olen on May 25 that Indianapolis's morning newspaper carried the headline, "FOUR-WHEEL DRIVE CARS LOOM AS '500' THREATS. "He was sending the "Oldfield" feature article to a list of 1000 newspapers throughout the country. A release "To the FWD Organization" announced that the company's entire Board of Directors would attend the race.

A separate internal release on this same date made the Organization aware of the following item in the Indianapolis Times of the 23rd: "Early Sunday Fred Merzney took the Coleman fourwheel drive car for a ride and cracked into No. 1 turn, demolishing the car. Sparky Sparks was riding with Fred, but neither was hurt. The car, however, was wrecked too badly and the Coleman people packed up and left for home."

The Coleman Motors Corporation of Littleton, Colorado also was a manufacturer of four-wheel drive trucks and thus a competitor of the Four Wheel Drive Auto Company. In 1930 they had entered two cars in the Indianapolis 500, both with front-wheel drive chassis of

Illustration 4-11 The Four Wheel Drive Company car with its authentic numbers and lettering. (RMRC)

Illustration 4-12 The "two's a crowd" William A.M. Burden entry with driver Gus Schrader, right, and mechanic Fred Blauvelt, left. (GBC ex IMS)

their own manufacture, powered by 4C 183 Miller marine engines. Driver Lou Moore crashed his car, but Phil Shafer went the distance, finishing a respectable 7th. The company did not compete the following year, but concentrated on the development of a four-wheel drive racing car, in which the two Miller fours were mounted side by side and were geared together. Trackside gossip in 1932 was that the car was not that badly damaged in its mishap, but was unqualifiable.

The Great Showdown

Harry Miller did not improve his luck by proclaiming his four-wheel drives the fastest cars in the field. True, his big V-8 engines combined most of the traits of the traditional thoroughbred with the displacement benefits of the stock-block formula. They could pump out around 300 bhp without the problems of supercharging, and four-wheel drive would deliver it efficiently to the pavement. They were virginally new machines; however; they had been built under extreme financial duress and had been deprived of any hint of running-in and development time. Getting them to the Speedway in time for qualifying trials had been a harrowing race in itself. Bob McDonough did very well to qualify his car 7th fastest out of a 40-car field. Dirt-track champ Gus Schrader, to whom Miller had entrusted the other four-wheel drive, qualified 11th, which was not bad for a new car and a driver who was new to the bricks.

Sources differ as to the lap on which Schrader smacked the wall; they range from 1st to 7th. Most give the cause of his mishap as a broken oil line. In any case, the $15,000 jewel was soundly wrecked on this truly fine driver's sole and ever so brief adventure at the Speedway. As for the Four Wheel Drive Auto Company car, McDonough lasted until somewhere between the 7th and 17th laps, depending on the published source, and he, too, had to drop out, "due to a broken oil line," but without bending his car. He, too, did not race at Indy again. He had competed there since 1924.

Four-wheel drive, for which so much had been claimed and from which so much had been expected, was a washout. It was barely mentioned in media coverage. Luck and fashion have much to do with mechanical evolution, and had four-wheel drive approached expectations at all, automotive history probably would have been markedly different. The psychological target had been gloriously missed, and it helped little that there would be other races down the line.

It was on the occasion of the 1932 Indianapolis 500 that the name of William A.M. Burden first became linked with Miller and four-wheel drive.

Chapter 5

A Four Wheel Drive Supercar for the Road

This machine and the concept behind it were in striking anticipation of the Bugatti EB 110 supercar of the early 1990s. It used 16 cylinders instead of the EB's 12, although a 45-degree V-8 version also was visualized. Both were to be Roots-blown, for immediate supercharger response throughout the rpm range. The V-16 was to pull around 500 bhp, against the EB's 600, both being more or less equally excessive for the road, although there was a lot of open road in 1932. And each was endowed with the advantages of four-wheel drive. Blasting down the straightaway and cornering on rails at high Gs in the EB is a stupendous experience. The Super Miller should have been just as out of this world, sixty years earlier.

The story is steeped in mystery. Out of nowhere, the name of William A.M. Burden appeared on the 1932 Indianapolis entry list, as entrant and implicit owner of the duplicate four-wheel drive racing car, which Harry Miller had been trying to sell to the Four Wheel Drive Auto Company up to the eve of the Indianapolis 500. Bill Burden was unknown at the Speedway prior to that race and would not be heard from there again. Who was he?

He was a great-great-grandson of railroad mogul Commodore Vanderbilt, the wealthiest man in the U.S.A. in his day. Bill was born in New York City in April 1906 and his father died when he was a baby. He took his B.A. from Harvard magna cum laude and went on to a life of astonishing achievement in many fields, including the financial, technological, cultural, artistic, and diplomatic fields in his country. Two small facts: For thirteen years he was director of New York's Museum of Modern Art and for twenty-three years chairman of the U.S. Government's Institute for Defense Analysis. His autobiography, *Peggy and I* (New York, 1982), is so modest that one has to turn to other sources for the facts in their overwhelming fullness. In his book he wrote, "Beginning when I was 12 I have owned 53 automobiles, small and large; the 1920s were the Golden Age for motors, and I thank my stars that I lived in a time when the finest in the world were being built." And he bought and enjoyed them. He had just turned twenty-six when he saw his Indianapolis entry fizzle out ignominiously in a smattering of laps. Thus began his adventure with Harry Miller.

My contact with Bill Burden came about through another remarkable achiever of approximately the same generation, Victor Emanuel. He was born into a very wealthy Dayton family in January 1896. On graduating from Dayton University in 1923 he became a partner in his father's utilities firm. By the early 1930s, he controlled a utilities empire that spanned twenty states and was valued by Fortune at $1.1 billion. Like Bill Burden, when the Model J Duesenberg arrived on the market, that's what he drove.

Leo Goossen told me how, in the summer of 1932, Emanuel had called Harry Miller to Manhattan, Victor Emanuel's then base of operations, for the purpose of discussing the design and construction of high-performance cars that, among other things, could account for themselves well in European competition. Harry Miller took Leo Goossen, his chief designer, with him. The encounter was to begin with dinner in Victor Emanuel's apartment. Leo said that, by the time they had penetrated this private palace within a palatial hotel, Harry Miller was mortally tongue-tied. As the liveried servants glided around them, it was Leo's lot to do all the talking for his boss. After they had safely gotten back to the real world, Harry Miller grumbled, "That man talks engines like a fool."

Acting on Leo's lead, I addressed a letter to Mr. Emanuel, who, in the intervening years, had replaced E. L. Cord as head of the multi-billion-dollar AVCO Corporation. Thus began an intermittent exchange of ideas between enthusiasts that lasted for years. This was his first letter:

October 20, 1954

Mr. Griffith Borgeson
454 South Maple Drive
Beverly Hills, California

Dear Mr. Borgeson:

I have your letter of the 14th. I knew Mr. Miller and admired his engines tremendously. I did give him an order for a car but it was never finished as his firm ran into financial difficulties and the trustees, or whoever took over its financial management, wanted an inordinately large amount of money to complete the car. I was going to have it road raced in Europe to show that for once an American car could win an important road race over there.

At the same time, Mr. William A. M. Burden of this city was having Mr. Miller build a car for him and my recollection is that that car was completed. If you should want to write him, his address is, care of William A. M. Burden Co., 75 West Street, New York 6.

I would very much like to secure a copy of anything you write about Mr. Miller if they are for sale.

Thanking you for writing me, and with kindest personal regards, I am,

Sincerely,

(signed)
Victor Emanuel

According to Chapter 57 of Dees*, Victor Emanuel had chosen the V-8 option in the supercar chassis. Guided by Mr. Emanuel, I eventually wrote to Mr. Burden. He, too, proved to be still the car enthusiast, and his response was frank and open:

*Mark L. Dees, *The Miller Dynasty, A Technical History of the Work of Harry A. Miller, His Associates and His Successors*, The Hippodrome Publishing Company, Moorpark, California, 1994.

November 5, 1954

Mr. Griffith Borgeson
454 South Maple Drive
Beverly Hills, California

Dear Mr. Borgeson:

Many thanks for your letter of October 22nd. It is true that I had a 16 cylinder, 4 wheel drive, 2-seater sports car built for me by Harry Miller. I have some photographs of the car which I will be willing to let you use.

It is difficult to give you full information about this machine without writing a book and I really haven't the time to do it. If you are ever in New York or on my next trip to California, I would be glad to talk to you about this.

I am sorry to say that the automobile was a complete failure and dealing with Miller on a project which he had not the technical or financial means to complete was extremely unsatisfactory.

In order to complete the car I had to take over the supervision of the small Miller shop myself. Fred Offenhauser, if he is alive, will doubtless remember the project in detail.

After many months delay and considerable expense, the car was finished and delivered. I ran it a few thousand miles but it was completely unsatisfactory. The chassis was too flexible, the gears were tremendously noisy, the steering column and steering assembly were poor and the steering geometry unsatisfactory. The car would have been dangerous to operate at really high speeds. I never ran it over 100 and then only for short distances.

I imagine you have already inquired about the V 8-passenger car which Miller made for Champion or Challenger in Santa Barbara some years before mine was completed. I don't know what luck he had with his but I hope it was more successful than mine.

If you would like more information, I would be glad to give it to you when you are in New York or when I am in California.

Sincerely,

(signed)
William A. M. Burden—

This and subsequent letters, plus photos, permitted me to write "The Ultimate Classic," which appeared in *Road & Track* in May 1955. To my knowledge this was the first article on this car ever to appear in print. As is my habit as a professional, before offering the article for publication I submitted it to its research source, Burden, for checking. This was his reply:

January 4, 1955

Mr. Griffith Borgeson
454 South Maple Drive
Beverly Hills, California

Dear Mr. Borgeson:

Many thanks for your letter of December 31st. I am glad you found the photographs excellent. However, I most decidedly do object to having a two-page pictorial feature on my car appear either in Road & Track or Motor Life with my name attached to it. You may have such an article appear provided my name is not used either with direct or indirect reference made in the course of the article. You could say the car was "built for a private owner".

I would appreciate your confirming this in writing.

Sincerely,

(signed)
William A. M. Burden

The strength of Bill Burden's feeling about not having his name associated with this project startled me. Many men make mistakes in life and laugh them off. There was no philosophical shrug or chuckle here. I don't feel that he had been traumatized by his experiences with Miller. I feel, rather, that here was a man with the strength of a force of nature who had gotten himself into one ridiculous situation after another. This did not fit the pattern of his habitual conduct. He would help me and the historical record, at the price of his own anonymity. I wrote further pieces about this car, always complying meticulously with this condition. But Burden had not made his wish clear to Goossen, who began to be interviewed with increasing frequency and through whom that fact eventually emerged. How Bill dealt with the story in his autobiography was to let his brother, Shirley, tell it. Shirley was in Hollywood, learning the film business, and working for RKO. He wrote:

> One day I received a telephone call from my brother in New York. He told me that he and Victor Emanuel, another car nut like himself, had ordered two Miller Special cars to be built in Los Angeles. He explained that Victor Emanuel had gotten tired of doling out money and seeing no car and had cancelled his order. That left my brother holding the bag. He reminded me of what he had ordered: a car of sixteen cylinders, four-wheel drive, 500 horsepower, supercharged. He wanted me to get a lawyer and go to the factory to see what was going on. I explained the situation to Dan O'Shea [head of RKO's legal department] and asked if he would help. A few days later Dan and I were on our way to look at the Miller Special. I don't know what Dan expected to see, but I visualized a small garage with a couple of greasy mechanics sweating over an engine block with holes in it. It took us quite a while to find the address, but the factory; and it was a factory in every sense of the word; took up at least three-quarters of a city block. Inside we met Fred Offenhauser, a charming gentleman in a white coat who was in charge of the operation.

The shop was filled with every kind of machine; lathes, drill presses, etc. At each machine was a mechanic working on a small, shiny something. Later, much later, when the trip through the plant was drawing to a close, Dan and I got up enough courage to ask Offenhauser where the car was. He explained in a tolerant voice [that] we had seen the car. All those shiny somethings when assembled would make one gorgeous Miller Special. As we left we asked how many people were on the payroll. The answer was twenty-five. We didn't talk much on the way home. There really wasn't much to talk about. The conclusions were obvious. These men were artists. Money was no object, especially when it belonged to someone else. If you wanted the car badly enough, you had to pay up on their terms. It was hard explaining this to my brother. I think he understood. A lot of time and a lot of money later, the Miller Special was finally finished. The torpedo-shaped body was painted black. It had two rather uncomfortable bucket seats, a low windshield, and no top. The engine, which seemed to take up at least three-quarters of the length of the car, was a joy to behold. Whether it ran or not mattered little. It should have been in a museum.

My romance with Flobelle [a niece of actor Douglas Fairbanks] was progressing, so I decided to take her for the first ride. It was summer. Large picture hats and diaphanous flower print dresses were in. There were very few houses on Cold Water Canyon Road in Los Angeles in those days. The grade was gentle and led to a beautiful view of San Fernando Valley at the top. I decided to take her there.

Flobelle was a sight to behold as she floated down the steps to Bill's black Miller Special chariot. The picture hat, the flower print dress, that beautiful blonde hair, and that face, that face, that wonderful face. The clothes were not exactly appropriate for the occasion, but who cared! I drove rather conventionally at first; I wanted to get the feel of the car. When we reached Cold Water Canyon I could not resist bearing down a little harder on the accelerator. The car leapt forward, as a true thoroughbred should. Flobe was getting a bit edgy. She didn't seem to understand why we had to go so fast even when I explained to her in a loud voice that I had to test the car in order to give my brother a full report on its performance.

I think we were going around eighty when it happened. There was a roar, the engine hood flew off and a geyser more spectacular than the one at Yellowstone National Park appeared from nowhere. I would have enjoyed the spectacle more had I not been so busy trying to stop the car. When I finally did, and took stock of the damage, I was speechless. Flobe, on the other hand, was not. She did everything but break off our engagement. Her picture hat was ruined, her dress was ruined, her shoes were ruined, her hair was a mess. I didn't look so good either, but that was my problem.

Several weeks later Mr. Offenhauser telephoned and apologized for the accident. He said there was nothing wrong with the car; the radiator needed adjusting.

I didn't drive the Miller Special much after that. I was too busy trying to get back into my future wife's good graces. The car was eventually shipped to Mount Kisco. I never heard from Mr. Offenhauser or my brother Bill why the radiator needed adjusting, but several years later a bird perched on my window ledge and told me when they finished building the engine it was too big to fit into the chassis, so they had to cut

> down the size of the radiator and the cooling fan. Result: There wasn't enough water to cool the engine.

Well, it made beautiful geysers, and gave work to twenty-five mechanics for a very long time. Bill Burden himself concluded the story with:

> I thanked Shirley for his help, but I was not reassured. When I tested the Miller Special at Mount Kisco I did so alone. The steering seemed to me dubious for so heavy a car and the higher the speed the more I distrusted the brakes. Peggy took a snapshot of me behind the wheel of my enormous black monster, which had cost $35,000. Eventually I sold it back to Fred Offenhauser for $600, the most costly disappointment of the many cars I have owned.

Illustration 5-1 Side view of the Burden super road car.

Illustration 5-2 Front view of the Burden super road car.

Illustration 5-3 Rear view of the Burden super road car. The year, barely visible on the license plates, is 1933. These dockside photos probably were taken at Los Angeles Harbor, just prior to shipment to New York. Like the other Miller four wheel drives, this one, also a victim of the Depression, knew no normal development. (William A.M. Burden, GBC)

Illustration 5-4 The bare chassis of what should have been the most stupendously performing road machine in history, up to and far beyond its time. Its similarity to the four wheel drive competition version was extreme, to its detriment, according to Offenhauser. (William A. M. Burden, GBC)

Illustration 5-5 Analysis of steering phenomena relative to location of king-pin axis; a very explicit document. Recess in backing plate suggests optimistic steering link. (William A.M. Burden, GBC)

Illustration 5-6 Bill and Peggy Burden in 1933, the year of the supercar. (William A.M. Burden)

Illustration 5-7 Bill's brother, Shirley, who took delivery of the super road car and gave it its initial road test. (William A.M. Burden)

Illustration 5-8 Bill and Peggy in attendance at the Easter Parade on New York City's Fifth Avenue around 1931. (William A.M. Burden)

Illustration 5-9 The only photo that we know of showing Bill with his four-wheel-drive Miller speedster. All brightwork except the hubcaps was painted the same black as the body. This may have been an effort to cool the car's ostentation in a Depression-ridden environment. This is a casual snapshot; if one cared about the car's appearance in the photo, the hood straps would have been buckled and the blemished paint near the rear fender would have been retouched. Note horizontal hood louvers. (William A.M. Burden)

Illustration 5-10 Coachmaker Rollston's rendering of the Model J Duesenberg that Burden drove from 1931 to 1939. Of all the coachbuilders who provided bodies for the Duesenberg J and SJ chassis, only Rollston used these horizontal louvers, with the exception of Bohman & Schwartz, who adopted them in about 1936. They may have been Bill's idea originally, since they also appeared on this Miller (William A.M. Burden)

Perhaps it was Offenhauser who acted as the fence in that transaction. It was Gerard Kirchhoff, who built the Chancellor body and worked with Miller on the Gulf Sixes, who in the late 1950s told me what had happened to the car. It was in 1937 that Miller himself had an intermediary buy the big speedster from Burden for, according to this source, $400. Initially, Harry Miller had the car trucked to the Gulf Oil plant in Pittsburgh, where it sat for several months. Then he had it hauled to the Gulf Research laboratories at Harmarville, Pennsylvania, where the Gulf four-wheel drives were under construction and development. There, he removed the aluminum body, which had been built by artist Emil Diedt. It was purchased by one of the lab employees. Harry Miller was interested only in the engine, which he shipped to a qualified shop in Indianapolis for thorough revision. He designed a new racing chassis to receive the engine, presumably using the drive-train components from the long, heavy road-car chassis. Kirchhoff himself worked on this spare-time racing-car project in '38 and '39. It never was completed, and I have yet to learn what became of the hardware.

If Bill, Shirley, and Flobelle thought they had it rough, they could have consoled themselves with the fact that they didn't have Miller's problems. He was blowing his fortune on this mad four-wheel drive gamble. Everything was going out and precious little was coming in; the dribbles from Four Wheel Drive Auto Company and Burden were not offsetting the outgo. His creditors were hounding him to death and, finally, in May 1933, there was nothing left to do but declare bankruptcy. Wiped out, crushed, he left Los Angeles, never to return alive. The Burden car was only about sixty-five percent complete. Offenhauser told me that Harry Miller had taken on the job originally "for around $35,000." There was a contract on which a lot of money still was due, and so it was arranged that Fred should be responsible for bringing the project to a proper conclusion. Burden really wanted his supercar and by the time he took delivery in Los Angeles, Fred said, he had had to put another fifteen or twenty thousand dollars into the operation, Harry Miller having miscalculated the real cost by that amount. "But Harry's biggest mistake," Fred said, "was that everything on that car followed racing practice. I always argued with him that you can't build a pleasure car by basing it on race-car principles. Sometimes he listened to Leo and me, but this wasn't one of them."

Fred was dead right in labeling this a pure racing car. The coachwork, drawn by Goossen, the lights, horn, and starter made it street-wise, but otherwise it was the Double De Dion track chassis, with a widened frame and the same choice of eight or sixteen-cylinder engines. In fact, both power units were hairier than the "pure racing" versions in that they were massively supercharged, and not by the sort of centrifugal pleasantry used on the SJ Duesenberg, but by a Roots blower as big around as the flywheel. It was placed between the crankcase and the flywheel, thus adding about a foot to the length of the power plant and pushing the car's occupants that much farther to the rear. As Miller historian Mark Dees judiciously points out, one of the blower's rotor shafts "carried the power right through from the engine to the clutch. This was a splendidly neat and original solution except for the length factor."

The fuel/air mixture was ducted from this blower through a manifold that lay within the 45-degree V formed by the cylinder blocks. It fed into horizontal inlet ports, as opposed to the siamesed downdrafts of the "pure racing" engine, new blocks having been designed and fabricated to permit this marvel of compactness. Here, the trade-off penalty was thermal: Instead of having an inter-cooler, one had an inter-heater, the manifold being smothered by hot blocks and valve-gear castings. No doubt, the same astute mechanical sleight-of-hand was to have been applied to the streets V-8, which never was built. As Offenhauser put it in his homely way, "Harry would always shoot the works on something tricky."

Descending to the basics, the bore, stroke, and displacement of the V-8 were 3.5 × 4, and 308 cubes, with about 500 bhp quoted at 5500 rpm. The same figures for the V-16 were 2.625 × 3.5, 303 cubes, and about the same output, but at 6000 rpm. Fred said that the Burden engine turned out to pull only about 400 bhp.

Wheelbase was a non-racing 130 in., about the same as the contemporary short-chassis Packard Twelve and far from the 142.5 and 153.5 of the J and SJ Duesenbergs. The Miller supercars were, however, aimed at the heart of that precise clientele. Quoted top speed with standard gearing was 140 mph for both models. Offenhauser assured me that constant-velocity U-joints were used in the Burden car, so they perhaps cannot be blamed for the handling problems reported by its owner, who pinpointed "unsatisfactory steering geometry." One hears the loud echo of Phillip Chancellor's complaint.

The Miller 91 cars and engines were the best that Miller ever built, because of the sheer quantity produced and the development that they underwent. This other practice of agonizingly producing Cellini-like one-off masterpieces was both unintentional and ruinous.

Automotive historian James D. Crank perceives the Burden adventure this way: "The Bugatti Royale was one man's statement concerning a supercar, made at a given moment in history. The Burden car was Miller's such statement. The Royale enjoyed the luxury of several years of reflective development before being released for symbolic 'production,' although the same engine, in railcar form, was made in good numbers. The Burden car was shot out of the sky before its first proper test run could be made. What impresses me profoundly is how advanced the concept was for its time. I see it as Miller's Royale."

Chapter 6

Miller Four Wheel Drive— An Official Description by the Builder

The following text appeared in *The Autocar* (London) for December 16, 1931. It was preceded by a long introduction by sports editor S.C.H. Davis, explaining to British readers a bit of what Miller cars were all about. Davis was well qualified to do that, having buddied with Count Lou Zborowski when that colorful figure was racing his Miller 122 at Brooklands track. Davis had also been Zborowski's riding mechanic in the car in the 1924 Grand Prix de l'ACF (Grand Prix de l'Automobile Club de France) at Lyon, France. The author of the text, Ted Miller, had his own qualifications: He was Harry Miller's son, and at that time sales manager of the company, in spite of what Preston Tucker was telling people about his role there. Since Ted, according to his mother, Edna, didn't know how to hold a screw-driver,* it's a safe bet that people such as Leo [Goossen], Fred [Offenhauser], and his Dad told him what to say, and that Harry Miller had to approve the text before anyone got to see it—which makes this an important, authentic document: It obviously was written before the cars had turned a wheel.

> Harry A. Miller has designed and constructed for competition on America's Speedways two racing cars which will be driven on all four wheels and which will handle better even than the front drive cars and provide complete wheel adhesion at all times.
>
> A better distribution of weight is possible with this type of car, inasmuch as the front and rear axle assemblies are of approximately the same weight. Also, the thorough tractive effort possible with a four wheel drive car affords the driver the ease of handling which heretofore has resulted in wheel slip or been forfeited in acceleration. These cars will be far more stable on the turns and will glide through an oil or water surface without slippage.
>
> These cars have but one clutch of the multi-disc type and a standard three forward speed and one reverse transmission. In this manner the power is applied to all wheels at the same time.
>
> The propeller-shaft drive consists of a cluster of gears, three in number, the driver, idler and driven, located at the rear of the transmission, and driving both the front and rear propeller-shafts situated on the right of the engine.

* Although not mechanically inclined, Ted was highly articulate and intelligent. He found his niche in life as the successful founder, editor, and publisher of specialized medical journals.

A differential cluster is located between the front and rear propeller-shafts in the propeller-shaft driving gear, allowing the front wheels to turn on a larger radius than the rear without locking. There is, however, a lock in the propeller-shaft differential which will drive the car if the front or rear driving axles should fail. This is not the main reason for locking the center differential out, it is to avoid wheel spin on low coefficient surfaces.

The front axle assembly is similar to Miller front drive cars, and conventional differentials are used in both front and rear, where the gear housing is bolted rigidly to the frame, eliminating unsprung weight, common in most cars.

A total of eight universal joints is used on the driving axles, six of which are the bronze block type with a working angle of 15 degrees, as used in all of the later type Miller race jobs, while the other two joints are the 30-degree angle double trunnion [constant velocity] type. These latter type are used as the front outboard joints, more commonly known as steering knuckle [hub carrier] joints.

The oil tank is located in the same unit with the front differential housing* for several reasons. This feature has been incorporated from past experience on the front drive racing cars, where it was found advantageous in cooling the oil; also it allows more weight to be placed on the front driving wheels. The radiator shell [and core] is mounted on top of this differential housing and the oil tank, and the frame side-members are also bolted thereto.

The chassis has cantilever [quarter-elliptic] springs for two tubular axles known as "dead axles" [De Dion tubes], and located directly in front of the forward driving axle and behind the rear driving axle.

Four wheel brakes are applied by means of cable, and can be operated by either hand or foot control. This cable feature has been characteristic of Miller race jobs since 1923.** The braking system was formerly operated by the hand lever for front and the foot pedal for the rear, but this is no longer permissible at Indianapolis.

Although applying the power to all four wheels, the wheelbase has been reduced to 100 in. in contrast to the 105 1/2 in. sister cars manufactured by Harry Miller for 1931, and is the same short length as used on the smaller single-seat jobs of four years ago. The 56 in. tread is retained from the past several years.

The latest designs are powered by a new type V-8 Miller engine with a bore of 3 1/2 in. and a 4 in. stroke, total piston displacement of 307.8 cubic inches (5045 cc), and supported on a three-point frame suspension.

Because of the high compression of these engines it is very difficult to crank by hand, and in order to comply with Indianapolis regulations which do not permit towing or pushing cars, these new jobs are equipped with 12-volt self-starters.

* "Differential housing" is an incorrect synonym for final-drive housing, which, in this case, is the transaxle housing, then known as "the front-drive unit."

**Miller pioneered Lockheed hydraulic brakes in 1919, found them to be imperfect, and returned to cable operation.

> Although Miller has been building eight-in-line engines for many years in all race jobs, he has chosen the V-type in these jobs as it permits a shorter wheelbase car and places more weight on the front wheels, an item which must be stressed and not overlooked.
>
> Sixteen valves are used, or two per cylinder, and the ignition system consists of two four-cylinder Bosch magnetos, while there are four downdraught new-type Miller carburetors placed on top of the cylinder blocks which are set 45 degrees apart, or 22 1/2 degrees off center.
>
> Five main bearings are used in this V-8 and the connecting rods are placed side by side and are similar to the 16-cylinder car made by Miller and driven at Indianapolis for the past several years.
>
> As the maximum piston displacement on the speedways and dirt tracks is now limited to 366 cubic inches (6000 cc) instead of the former 91 1/2 cubic inches, larger power plants developing far more horse-power are, of course, constructed which naturally result in proportionate wheel slip. By using four wheel drive, the wheel slip is reduced to a minimum, and it is Harry Miller's opinion that the near future will see the higher-powered cars all built with four wheel drive.

The concluding statement makes one wonder how the clairvoyant Miller, seconded by his go-getting associate, Tucker, managed not to perceive the rumblings of the all-wheel drive revolution in the military field. Symbolized by the Jeep, this revolution was fermenting just around the corner of the probable. Ettore Bugatti saw it coming, but got derailed by rail-cars; Art Herrington saw it coming and scored all the way.

Illustration 6-1 What is visible here pertains to 1932 as well as 1933, except for the fuel tank. The 45-degree V-8 engine endured both years, painfully. Both front and rear suspension are shown here to good effect. (RMRC)

Illustration 6-2 In this closer view of the beautiful 4 cam, 308 cu. in. engine one can see clearly how the Miller downdrafts bolt directly to the upper block surface. The inlet manifold log was an internal element of the block/ head casting. (RMRC)

Illustration 6-3 The 308 V-8 marked a tuning point in Harry Miller's engine design career in that he gave up the barrel crankcase for the split type. But the split was far below the crankshaft centerline, the main-bearing supports being integral with the crankcase. (RMRC)

Illustration 6-4 A 1932 rear-end view, showing that year's fuel tank and location of the final drive housing and live axle shafts. (GBC ex IMS)

Illustration 6-5 A fascinating document: Miller's answer to what a transfer case should be. This design contained two main gears, which, herringbone-cut or not, pumped awful heat because their housing was too tight. (GBC ex IMS)

Illustration 6-6 A photo that shows a front spring being shackled to its De Dion before installation of snubbing blocks, which were intended to replace shock absorbers. (FAUC)

Illustration 6-7 The brightly chromed snubbing block which was supposed to act as a shock absorber. Note the superbly machined finish of the De Dion tube. (GBC ex IMS)

Illustration 6-8 An eloquent photo of the Four Wheel Drive Auto Company chassis at the Cornell Aeronautical Laboratory Flight Research Department while being prepared for Pike's Peak 1948 (see Chapter 18). In particular, the gear ratios were being changed so they would be more suitable for Pike's Peak. Prominent here are rear frame kickups, paltry Houdaille shock absorber, and rear-axle assembly cut loose from its spring supports. (RMRC ex WFMC)

Illustration 6-9 This photo was taken at the same photo session at Cornell. It presents a good view of the husky hydraulic shock absorbers which had been added to dampen the front springs under hill-climb and road-racing conditions. (RMRC ex WFMC)

Chapter 7

1932 Post-Indy Repercussions

The first bit of hard data to surface after the race was a table that purported to show the rise in air pressure of each of the four tires of representative front-wheel drive, rear-wheel drive, and four-wheel drive cars at the end of twenty-five mile qualifying runs (Illustration 7-1).

At Start		At end of 25-mile qualifying run			
		Right rear	Left rear	Right front	Left front
Front drives	38	43	41	45	44
Rear drives	38	45	44	43	41
FWD	38	42	41.5	42	41

Illustration 7-1 Tire pressure changes.

This is a very interesting bit of reference material. The car's sponsor interpreted it thus:

"The ... chart shows that the four-wheel drive with its uniform distribution of load and power results in a lower and more uniform temperature in all the tires which means lower pressure, less heat and less strain." The figures of course represent pressure, not temperature.

At some time prior to June 3rd an outside source had proposed that the Four Wheel Drive Auto Company car should take part in an upcoming one hundred-mile race on the Detroit one-mile dirt track. Olen asked Engel, in Indianapolis, to check on the feasibility of this. Engel replied with a telegram, which is missing from the file, but Olen's reply makes the contents clear. Unfortunately, the contract referred to in Olen's letter is also missing from the file.

June 3, 1932.

Mr. H. G. Engel,
532 E. 54th St.,
Indianapolis, Indiana.

My dear Mr. Engel:-

We have your telegram stating it would not be possible to have the car in the race at Detroit for which I am pleased.

Will you please inform Ernie Olson and those in charge of the car, that it is our wish that the car be placed in A#1 condition, of course, under the supervision of Harry Miller. When the car is in proper operating condition, then we desire to have it taken to Purdue University for them to use for test as they desire for a period of at least three days. We will give you and Ernie Olson further instructions at that time.

I wish that you would tell Mr. Olson that under our contract with Harry Miller Inc., the car is to be put in A#1 condition and for this reason we do not want him to make any changes in the car of any kind that are not authorized by Harry Miller.

After we have finally accepted the car we will be glad to talk over with him anything that might at that time lead to a further improvement.

The enclosed letter to Ernie Olson will clear up the question of just who is in charge of the matter to your satisfaction and also to his.

Upon returning here and conferring with my Board of Directors it was the concensus of opinion not to put the car on a dirt track for the time being. We all desire to have the car placed in A#1 condition and be sure everything is right before we attempt any further racing.

All of us feel that we would like to get acquainted with Bryan Saulpaugh as the possible future driver for the car in case we take it to plant and dirt tracks.

In the conference I had with the drivers and mechanics in my room I requested each one to write a statement of what they found to be the outstanding things in the car. I wish you would try and gather up these statements. We intend to publish a little booklet giving these statements, also a picture of each one of the drivers and mechanics together with a picture of the car.

Yours very truly,
FOUR WHEEL DRIVE AUTO CO.

W.A.Olen, President
EL

Ernie Olson's post-race responsibilities had not been clearly defined, which Olen took care of, also on June 3rd.

June 3, 1932.

Mr. Ernie Olson:

In order to avoid any confusion we will send any instructions that we may have for you to Mr. Engel as long as the car is at Indianapolis.

I have written Mr. Engel that when the car is in proper operating condition to your entire satisfaction and you have reported to us and we approved it from here, then we shall want the car taken to Purdue University there to remain for a period of at least three days and to be subject to such trial and inspection as the authorities at Purdue may require.

I trust this will be entirely satisfactory.

Yours very truly,
FOUR WHEEL DRIVE AUTO CO.

W.A.Olen, President
EL

On June 5th, Bryan Saulpaugh was the first of many drivers who, recognizing the virtues of Miller four-wheel drive, sought a place on the company team. He had driven the Miller sixteen-cylinder car as a rookie at Indianapolis in 1932, and was killed in a race on the Oakland Speedway the following April.

Although the Four Wheel Drive Auto Company car had not even been properly run in, when Olson opened up the engine he found its lower-end bearings to be wiped out. Having reported this to Miller, he wired Olen.

GU7 60 DL 2 EXTRA XU=AU INDIANAPOLIS IND JUNE 6 1932 1044A

W OLEN=

F W D CO CLINTONVILLE WIS=

ALL MOTOR BEARINGS BAD SHAPE MILLER WANTS ALLISON BEARINGS THROUGH OUT APPROXIMATE COST OF MAIN AND ROD BEARINGS SIX HUNDRED FIFTY DOLLARS WILL YOU AUTHORIZE SPENDING THIS SUM IF SO WIRE A DEPOSIT ON THIS ORDER I BELIEVE IT IS VERY IMPORTANT THAT WE INSTALL THESE BEARINGS PLEASE HAVE MR GRAVES SEND ME A GOOD SMALL OIL COOLER=

ERNIE OLSON MERZ ENGINEERING.

1121A.

In 1932 $650 would buy you a pretty good automobile, and Olen was aghast. He used Olson to impress upon Miller the responsibilities of the parties to the contract.

June 6, 1932.

ERNIE OLSON
MERZ ENGINEERING
INDIANAPOLIS INDIANA

HARRY MILLER CONTRACTED TO FURNISH US A RACING CAR SUITABLE FOR THE INDIANAPOLIS RACE WE MUST HOLD HIM TO THE COMPLETION OF HIS CONTRACT THIS REASON CANNOT AUTHORIZE ALLISON BEARINGS IT IS ENTIRELY UP TO MILLER

OLEN

WAO:ELV

Still, on June 6th, Olen, a legal mind, spelled things out to Engel. Olen did not know that the design was defective very largely because Harry Miller was expected to build the car on faith, rather than with coin, with the result that there was not a moment left for development, in the course of which the bearings and other imperfect solutions to the task could have been rectified. As for Barney Oldfield's leftover change, good luck!

June 6, 1932

Mr. H. G. Engel,
532 E 54th St.,
Indianapolis, Ind.

My dear Mr. Engel:-

I am enclosing copy of a telegram from Ernie Olson, together with our reply.

I wish you would take up with Ernie Olson at once and tell him that we are not going to make any alterations in the car whatever. Our contract with Harry Miller provided that he was to build us a car suitable for the Indianapolis race. He has not furnished such a car. We do not want any divided responsibility, but, on the contrary, we want to hold Mr. Miller and give him an opportunity to furnish the kind of car he agreed to. If bearings are defective, it is up to Mr. Miller to furnish proper bearings, or anything else that is wrong with the car.

Please instruct Mr. Olson not to make any repairs on the car that have not been authorized by Mr. Miller in writing.

I wish you would give me a full report on just what condition you find the car to be in. Have you had any talk with Miller about it? Your letter was so extremely short that we were not able to get much from it. You did not say with whom you had talked and that is very important.

The check for $500.00 was endorsed over to Barney Oldfield with the understanding he was to pay for all gasoline, oil, tires and other supplies that were bought for the race. This was a part of their contract anyhow. There was about $180.00 due us which we turned over to him.

We trust you will have patience to clear this matter up. In order that there may be no misunderstanding, we want everything to pass through you on this car.

Yours very truly,
FOUR WHEEL DRIVE AUTO CO.

W.A. Olen, President
EL
Encl.

On the 7th, Engel reported to Olen with infinite patience. Olson, who was one of the greatest racing machinists and mechanics of all time, who rode with Jimmy Murphy, and upon whom Frank Lockhart and other greats depended, was being paid a pittance here. On top of that, his position was ambiguous, yet he was being used as the executive intermediary between these two industrial chiefs, Olen and Miller. Engel saw both sides of the problem and was feeling the strain of also being a buffer. While $32.58 was a lot to pay for a set of final-drive gears, the exposure at Indianapolis had stirred up the procurement section of the State Highway Department.

To Mr. W.A. Olen at Clintonville, Wisconsin

From H.G. Engel at Indianapolis, Indiana

June 7, 1932

Dear Mr. Olen:

I have received your letter regarding the race car and will follow out your instructions.

I delivered your letter to Ernie Olsen last evening and had quite a talk with him. I explained to Olsen that we did not know anything about the racing business and that was one of the reasons that we contracted with Harry Miller and that our contract with Miller called for a race car complete for a certain sum of money and that there was to be no additional expenses whatsoever except the salary of one mechanic.

Olsen talked with Miller on the telephone last night and Miller said it was not up to him to replace any bearings, but that he thought he would call you on the telephone.

Some of the babitt bearings will need replacing although they are not burned out. Olsen stated he can replace these with babitt bearings for around $200.00. He explained to me that it is sometimes necessary to replace bearings after every race because of the terrific heat that they are put to and that if we would replace them all with the Allison bearings this time it would cost around $680 that he feels we would be money ahead in the long run. This is also Mr. Millers suggestion that we put in these Allison bearings in order to save money in the long run. As I understand it this Allison bearing is something new.

In regard to the additional gear ratios Olsen informs me that it is necessary to have a number of different gear ratios for any race car. For example they used one for qualifying and a different one for the race here in Indianapolis, but when the car is used on a dirt track it is necessary to have still another gear ratio. Mr. Miller ordered the 3rd gear ratio from the Chrysler people which would be suitable for use on a mile dirt track. We have not picked this up as yet. The cost is $32.58 and in addition to that there will be some machine work that will have to be done on these before they can be put in the car. Mr. Olsen informs me that a race car usually carries from six to ten different gear ratios with them as spare parts when they travel from place to place racing.

This State Highway deal in Indianapolis continues to look good and we expect to have a decision Wednesday or Thursday of this week. Mr. Borne and Mr. Busch both told us that the deal looked good for six to ten trucks they could not tell sure until the district engineers all made and got checking into their budgets. I feel that this is a step into the right direction as all of these trucks will go into the southern part of the state for all purposes highway work. Some of them are equipped with under body scrapers. This is one of the things we have been trying to put over for three years and I feel sure that once we get started that we will continue to receive a good deal of the business each year on four wheel driven specifications.

Very truly yours,

H.G. Engel
HGE/PEC
[handwritten post script]
P.S. I understand Harry Miller will be in Indianapolis the end of the week.

A new development caused Engel to leapfrog his letter of the 7th to Olen with a night-letter telegram.

GU1 77 NL COLLECT= INDIANAPOLIS IND JUN 7 1932

FOUR WHEEL DRIVE AUTO CO=

CLINTONVILLE WIS=

OLEN STOP MILLER EXPECTS TO BE INDIANAPOLIS THURSDAY OR FRIDAY STOP SUGGEST YOU SEND ME COPY OF CONTRACT AND CORRESPONDENCE SO THAT I CAN TALK OVER ENTIRE DEAL WITH HIM STOP OLSON HAS MADE NO CHANGES EXCEPT THOSE AUTHORIZED BY MILLER STOP NO MAJOR CHANGES HAVE BEEN SUGGESTED BY MILLER EXCEPT NEW TRANSFER CASE BEARINGS OIL COOLER STOP OLSON HAS SEVERAL SUGGESTED CHANGES BUT I ADVISED HIM TODAY NOT TO MAKE THESE UNLESS AUTHORIZED BY MILLER=

H G ENGEL

JUN 8 807A.

Olin replied on the 8th with a very terse wire. The poorly legible carbon copy in the file reads: "COST OF ADDITIONAL GEAR RATIOS OKAY DESIRE YOU WAIT TO SEE MILLER ON BEARINGS THEN REPORT". He ignored Engel's request for a copy of the contract.

Engel, his request having been snubbed, leaving him to conduct himself in a responsible manner without knowing the rules of the game, replied to Olen's wire with a calm summary of the facts. The "Mr. [Frank] Gause" referred to the person who held the title of Secretary, Four Wheel Auto Drive Company.

To Mr. W.A. Olen at Clintonville, Wis.
From H.G. Engel at Indianapolis, Indiana

June 8, 1932

Dear Mr. Olen:

In regards to my telegram I had no intention of entering into a controversy with Mr. Harry Miller. My thougt was to explain to Mr. Miller that we contracted with him for the furnishing of a car and the handling of it at the race because we were unfamiliar with the racing business.

I thought that I should have a copy of the contract so that I would know just what it contained. It is my understanding that the contract states that we are to be put to no additional expenses whatsoever except the salary of one mechanic.

The way the thing was handled it seems that there were a lot of expenses that Miller feels that we should pay. In telephone conversation last night Mr. Miller stated that he had completed his contract, that he had furnished us a car that met the requirements and any trouble that developed after that was up to the Four Wheel Drive Auto Company to take care of.

He states that this is the usual practice in furnishing a race car. That even though they find changes have to be made that it is up to the purchaser to make them.

When I talked to Mr. Miller on June 3rd he told me that he recommended that the present bearings be replaced with Allison bearings and that he had ordered a set of these bearings for our car. He did not say at that time that he expected us to pay for them and I did not question him as I assumed that he was going to pay for them.

I wrote you on June 3rd, stateing that Mr.Miller had ordered these bearings from Allison. The Allison people got in touch with Mr. Olsen and told him they would not fill the order until a deposite had been made and that was the reasons fcr Mr. Olsen's telegram to you.

I do not know what your arrangements with Mr. Oldfield was regarding the payment of bills, but he evidently left town without taking care of some of them.

Gasoline is furnished free by the Richfield Oil Co. Mr. Miller did not recommend the use of oil that was furnished free at the speedway and that is the reason that some was purchased.

Mr. Olsen seems to be working in accordance with your wishes and is making no changes except those recommended by Mr. Miller. The car is completely torn down. It seems that this is the usual practise after a race to inspect every working part and to be sure that they are put back in A1 condition for the next race. They are at a standstill now as there is nothing more they can do until they get authority from someone to go ahead on bearings. I told Mr. Olsen that I did not authorize the expenditure of any money outside of small incidentals like solder etc.

I sent into Mr. Gause a complete list of the expenditures that had been made along with receipts for same.

As per our conversation of this morning Mr. Miller expects to come to Clintonville and talk things over with you the end of this week or the first of next. Trusting that my actions met with your approval I am

Very truly yours,

HGEngel

H.G. Engel
HGE/PEC

The Gause memo just mentioned suggests Engel's own meticulous business practice.

To Mr. Frank Gause at Clintonville, Wis.
From H.G. Engel at Indianapolis, Indiana

June 8

Dear Frank:

Mr. Olen authorized me by wire to pick up a set of gears from the Carl H. Wallerich Co. to be used in connection with our race car and to pay for a set of gears that was ordered by Mr. Miller.

I am inclosing a receipt for $69.32 covering this expenditure and will appreciate your depositing this to my account at the bank.

Very truly yours,

H.G. Engel

H.G. Engel
HGE/PEC

On the 9th Engel reported to Olen on Harry Miller's visit of the previous evening, doing so with his usual calm lucidity and objectivity. What is particularly interesting in this letter, is that the need for a transfer case of different design already had been recognized. This was the downfall of the two cars as they ran at Indy in 1932. Again, had a minimum of development time been available, the awful defect could have been spotted and possibly corrected in time. In May 1954 Leo Goossen told me, matter-of-factly, "I goofed on the final-drive [transfer case] design. I kept the housing so close to the gears that they acted as a pump. They generated terrific heat and pressure and finally the lube oil squirted out of every joint. The metal turned blue from the heat. The Four Wheel Auto Drive Company re-designed their transfer case, using three or four gears instead of our two, and they had little trouble, if any."

Obviously, Olen's specification that Miller was to "use our type of center differential" was not put into effect. Miller may have argued that the design was too heavy for racing practice and did it his way with the client's consent. Taking this mistaken fork in the road may well have changed the course of automotive history.

The spectacular loss of oil by both four-wheel drive cars during the race provided the basis for the theory that a "broken oil line" was the cause of their early retirements. In the 1994 booklet of the Four Wheel Drive Auto Company Museum this is still the given reason.

To Mr. W.A . Olen at Clintonville, Wisconsin
From H.G. Engel at Indianapolis, Indiana

June 9

Attention Mr. W.A. Olen

Dear Mr. Olen:

Mr. Miller came over to Indianapolis last night to look over the car and to talk with me. He had expected to be able to come up to Clintonville this week and talk the proposition over with you, but he has some additional business in Dayton, Akron New York and Detroit that require immediate attention and he will be unable to come to Clintonville this week. He plans to see you before he goes back to the coast which will be approximately ten days.

Mr. Miller states that the bearings on our race car are not defective that the reason for them burning out was because of lack of oil. He admits that there was something wrong with the oil system and that he is correcting that trouble free of charge to us.

He explains that the duties of a riding mechanic is to watch the oil pressure and at the time Billy Arnold drove this car the riding mechanic was watching the driver and not the oil pressure and that if he had been watching the oil pressure and would have signaled the driver to stop the bearings would not have burned out and because he rode for some time with no oil pressure it caused the other bearings to flake out and that this is no fault of his.

He at first recommended Allison bearings but he feels now that the cost of these bearings are out of line and that they should be replaced with standard babbit bearings.

Mr. Miller is willing to cooperate with us and to do his share, and as stated above is correcting the oil trouble free of charge and if we will send these bearings back to Los Angeles that he will replace them free of charge, but he does not feel that we should ask him to pay for having bearings replaced at Indianapolis. It will take approximately ten days to get the bearings back from Los Angeles.

As explained in previous letters Mr. Miller is also going to replace the transfer case with a larger case and different design free of charge.

Mr. Miller states that it will also be necessary to have the crank shaft re-ground do to the bearing burning out when Arnold drove the car which was no fault of Millers and that he feels we should stand this expense which would be approximately $50.00

Mr. Miller explained to me that the usual practise in buying a race car that all you purchase price included is the car and that [it] is usually up to the owner to purchase the tires and other miscellaneous equipment.

He stated that he had done more for us than he ever has for anyone else. He paid for the tires that we used on our car in the race. In regard to the extra set of wheels Mr. Miller feels that this is necessary spare parts and that we should pay for them however, if you do not desire this extra set of wheels he will take them back. The cost of these wheels are included in an itemized expense account along with receipt that I sent to Mr. Gause on June 3rd, total of which is $121.55 which also includes some miscellaneous items.

Mr. Miller wanted me to explain to you that if we desire to give this car the 100 mile test before it leaves the track that the expenses of such a test will be ours. In order to make this test we will have to hire a driver, buy some tires and oil. Gasoline will be furnished free of charge by the Richfield Oil Company and the Speedway Corporation will make no charges for the use of the track and Ernie Olsen will act as riding mechanic to watch the oil pressure and tires and perform the regular duties required.

I am quoting my conversation with Mr. Miller as near as possible. Would appreciate you giving me instructions on this as soon as possible as the mechanics are more or less at a standstill after Saturday and they are on our pay roll. It would take approximately 3 1/2 days to have the bearings done here at a cost of from $150 to $200.

Yours very truly,

H.G., Engel
HGE/PEC

Among the services rendered by publicist Herman Deupree was this harvest of expert post-race opinion. The diagnoses of "What Had Gone Wrong" were as wide of the mark as they were confident. The stationery is that of FWDCO's Indianapolis agent, who had provided temporary office space.

June 9, 1932

Four Wheel Drive Auto Company
Clintonville, Wis.

Gentlemen:

Attention Mr. W.A. Olen, President

The writer promised to interview a number of persons who have had experience with the racing car business and render you a report. While I have talked to a number of persons I can get all of the various ideas that were brought out from the ideas of four different persons.

These men are as follows:

Wm. F. Sturm, newspaper and national magazine writer and the American Manager for Major Seagrave, Sir Douglas Campbell and Kaye Don, the Englishmen who broke the world's records at Daytona Beach.

Carl H. Wallerich, Chrysler Distributor and who has been an active official judge of the 500-Mile Race for many years.

Wm. Guy Wall, who needs no introduction to you.

Louis Schwitzer, of the Schwitzer Cummins Company, a very brilliant engineer who has been adviser to many companies in the racing business, who has been American representative of the Mercedes Company in their American activities and has been the head of the three A Technical committee at Indianapolis and elsewhere for many years and who also has a very close personal interest in the well-fare of the FWD Company, because as I understand him, you have been a good customer of his firm and he has a high personal regard for you, your company and your product.

Sturm's idea is as follows:

"Get the car in shape and go after records, either on the Speedway or the Beach." He outlined this plan to you the evening of the drivers' meeting.

(Many persons including myself want to bring up this point-that such private trials of themselves do NOT attract much publicity. To get the message to the public it is necessary to buy many thousands of dollars worth of newspaper and magazine space.)

Col. Wall's expression is this:

"Get the car in shape and race it all over the country." He says that the four-wheel-drive principle of power application was demonstrated by the Miller-FWDs to be superior to either front or rear-wheel jobs on any kind of a race track, brick, dirt or board. He also feels that you would get a world of publicity. He urged that you enter the car in next year's race by all means-if it took then to get the motor worked out.

This opinion of Mr. Wall's was also expressed by Bert Dinghley, former Stutz Official and now vice-president of the Marmon-Harrington Co. His idea would be to carry the race car on an FWD truck, with flaring banners telling all who see it about the FOUR WHEEL DRIVE.

Car Wallerich says:

"If the bugs can be worked out of the motor the job should be brought back next year because nothing else in the race could compete with it because of its ability to negotiate the curves at high speed."

(Now this respect for the superiority of the four-wheel-drive principle in racing is really universal among the entire racing fraternity, including drivers, mechanics, engineers, etc.)

Mr. Louis Schwitzer, went into some detail which I will try to condense as follows, quoting verbatim as near as possible:

"Harry Miller is a wonderful and brilliant designer of racing cars, WITHOUT the fundamental schooling and training to work out many important engineering problems. I have great respect for him, but know his limitations after checking his jobs for many years.

"I felt that his oiling system was fundamentally wrong on my first inspection of the job. There is a serious problem in all opposed motors and many leading engineers have spent weary months trying to get proper oiling on such types of motors."

"The oil syphons out of the Miller-FWD jobs as fast, almost as you pour in it. Harry Miller cannot correct this defect without help. I certainly would get that FWD Car to the FWD Factory. They have capable engineers there, proper tools etc., to put the job in shape.

"Personally I have never felt that the worries and grief of the race-car business was worth what it cost to a big manufacturer. I would much prefer to put the car in shape and sell it to someone of the Hartz type, who makes a business of racing.

By giving such a promoter a comparatively low price it could be in the contract that it should be called the FWD and the factory would get all of the attending publicity with none of the worries and expense of promotion and would not be mixed up with the racing fraternity generally whose wild ideas as a rule are enough to distract a sound business man at the head of a big manufacturing plant.

"There is absolutely no question about the merit of the four-wheel-drive principle. The superiority of this principle as applied to speed cars was well demonstrated in the trial speeds efforts of the two new Miller jobs."

"I, also, believe a big standard FWD Truck with a lot of chromium, nice big hub-caps and a distinctive paint job loaded with a pay load and let Cannonball Baker take it from New York to Los Angeles. He will break all records and get a world of valuable publicity based on PAY LOAD figures that will actually appeal directly to the motor transport trade which should know about and be interested in the superiority of the four-wheel-drive principle."

"I would not spend a lot of money for new bearings in the race job until the oil trouble was corrected. The problem in the job is not bearings but Oil-supply."

(Now this diagnosis of the car is absolutely supported by Don Herr, a great racing driver and owner of the largest automobile maintenance shop in Indianapolis. Also by Carl Kaiser, a notable success in the racing car field, an expert in motor rebuilding and a very fine practical mechanic.)

The above report covers a canvass of persons of successful experience and intelligence.

The writer would also like to take this opportunity to tell you how much he enjoyed his association with you during the race here.

If there is anything that I can do to be of service to you please know your command will be happily executed to the best of my ability.

Yours very truly,

Herman G. Deupree

H.G. Deupree

On June 10th, Engel decided to wire to Olen certain essentials of his letter of the previous day.

GU11 55 DL COLLECT=INDIANAPOLIS IND JUN 10 1932 124 1P

THE FOUR WHEEL DRIVE CO=

CLINTONVILLE WIS=

OLEN MILLER UNABLE COME CLINTONVILLE AT THIS TIME STOP CLAIMS BEARINGS NOT DEFECTIVE BUT BURNED BECAUSE OF FAULT RIDING MECHANIC NOT WATCHING OIL PRESSURE STOP HE WILL REPLACE BEARINGS IF WE SEND THEM HIS FACTORY STOP WILL ALSO CORRECT OIL TROUBLE WE TO PAY FOR REGRINDING CRANKSHAFT STOP COMPLETE DETAILS FOLLOWS IN LETTER=

H G ENGEL.

111P.

On the 7th, a pair of connecting rods and a pair of valves for the race car arrived at Clintonville, from Los Angeles. A Receiving Report in the company files documents this. Perhaps Ernie Olson had told Harry Miller that they were needed. Chief Engineer Dodge was in the dark and asked Engel.

June 7, 1932

Mr. H. G. Engels
532 E. 54th Street
Indianapolis, Ind.

Dear Bud:

We just received two connecting rods and two valves for our racing car.

If you need these parts, kindly wire us and we will ship them to you; but unless we hear from you, we will hold same in our stock here.

Yours very truly,
FOUR WHEEL DRIVE AUTO CO.

H. B. Dodge,
Chief Engineer.

LAN

These two items are trivial, but help to clarify this most interesting night letter from Engel, which is earmarked for no one. Miller planned to take his own four-wheel-drive car to Argentina! International road racing was in his mind.

GU3 50 NL COLLECT XU=INDIANAPOLIS IND JUN 10 1932

FOUR WHEEL DRIVE SALES CO=

CLINTONVILLE WIS=

PLEASE WIRE ME CODE ADDRESS OF BUENASAIRES DEALER AS MILLER WANTS TO CABLE HIM REGARDING RATES AS HE PLANS ON TAKING HIS CAR THERE STOP RETURN CONNECTING RODS TO LOSANGELES THEY WILL HAVE TO BE CHANGED DO YOU AUTHORIZE WORK DONE ON CRANKSHAFT PAID BY US APPROXIMATELY FIFTY DOLLARS=

H G ENGEL.

809A JUN 11

On the 8th Pop [T.E.] Myers sent to the Four Wheel Drive Auto Company a list of its car's official lap times, establishing the fact that the machine completed six laps.

June 8, 1932.

Four Wheel Drive Auto Co.,
Clintonville, Wis.

Gentlemen:

Attached you will find a tabulation made by Mr. Porter, the Chief Timer, giving the elapsed and total time, as well as the miles per hour for each of the six laps covered by your car No. 58.

We trust that this will be of service to you and are very glad to send you this information.

Very truly yours,

INDIANAPOLIS SPEEDWAY CORPORATION.

T.E. Myers
Vice-President.
TEM:hr

The list, as transmitted to Olen by Hanson, follows.

June 10, 1932

Mr. Olen:-

Following is an account of the mileage record of our car at the race in Indianapolis, as supplied by Mr. Odis Porter, Chief Timer.

Very truly yours,

W.M. Hanson
Advertising Manager
PGS

CAR	LAP	ELAPSE TIME	TOTAL TIME	M. P. H. PER LAP	TOTAL AVERAGE FOR SIX LAPS
58	1	1:40.28	1:40.28	89.748 +	
58	2	1:25.07	3:05.35	105.795 +	
58	3	1:24.89	4:30.24	106.019 +	
58	4	1:24.67	5:54.91	106.295 +	
58	5	1:24.76	7:19.67	106.182 +	
58	6	1:33.29	8:52.96	96.478 +	In Pit 101.302 +

On June 11th, Engel wrote to Olen concerning Harry Miller's desire to race both four-wheel-drive cars in South America. The lately discovered need to regrind the crankshaft rendered the replacement connecting rods unusable unless rebabbitted and rebored.

Date 6/11/32

To Mr. W.A. Olen at Clintonville, Wisconsin
From H.G. Engel at Indianapolis, Indiana

Dear Mr. Olen

Harry Miller contemplates taking his FWD race car to South America and when he gets to Clintonville is going to ask you to allow him to take our car along with him. He [wishes] to race the two of them as a team provided he can get the proper guarantee and he requested that I wire you asking for the code address of our South American dealer as he wants to cable him to look up certain information pertaining to the race. If we do not wish to enter our car along with his and the guarantee is right he is going to send his car anyway.

Mr. Miller has sent the connecting rods and main bearings back to his place for replacement and it will be necessary to get the crank shaft reground before they can begin work in Calif. as the regrinding will change the dimensions of the bearings it will be necessary for us to send the two rods that we have in Clintonville back to Los Angeles because they will not fit after the shaft has been reground.

Very truly yours,

H.G.Engel
HGE/PEC

Someone, most likely Advertising Manager Hanson, had the idea of obtaining testimonial letters, apparently free, from everyone even remotely qualified to praise the four-wheel-drive car. So Engel had been put to work on this. The "Mr. Kriss" of whom he speaks is Pete Kreis.

6/11/32

To Mr. W.A. Olen at Clintonville, Wis.
From H.G. Engel at Indianapolis, Indiana

Dear Mr. Olen:

I am attaching a letter from Mr. McDonogh regarding the handling of our car at the Speedway.

I am having considerable difficulty in getting these boys lined up as most of them have left town. Mr. Kriss told me he would be glad to give us a letter, but he is working for the Studebaker corporation and thought he should get permission from them to do this.

I have written letters to all the men that drove the car, at their home address, but have not received any replies as yet. I will also get one from Ernie Olsen regarding the view point of a mechanic.

Very truly yours,

H.G. Engel
HGE/PEC

The first testimonial, that of Bob McDonogh, was again on General Equipment stationery, inappropriately.

GENERAL EQUIPMENT & SUPPLY COMPANY
31 W. Eleventh Street,
Indianapolis, Indiana.

June 11, 1932

Four Wheel Drive Auto Co.
Clintonville, Wisconsin

Attention Mr. W.A. Olen

Dear Mr. Olen

I would like to express myself regarding the actions of the Miller FWD Special on the track at Indianapolis. I found absolutely no slippage in going into a turn and pulling through as compared with a rear drive or a front drive at comparable speeds.

As you know there are some rough spots on the Indianapolis track and I found that there was absolutely no slipping as experienced in other types of drives.

I had the car out one day during a rain and found that it handled the same on a wet track as well as a dry. In both cases we got up to a straight away speed of 125 miles an hour. I appreciated the opportunity of driving the car and enjoyed the experience very much and if you decide to race this car next year I would like to put my bid in for it.

Very truly yours,

Robert W. McDonogh

Also on June 11th, Engel wrote to Gause concerning matters dealt with in a letter of his of the 6th, which we do not have.

6/11/32

To Mr. Frank Gause at Clintonville, Wis.
From H.G. Engel at Indianapolis, Indiana

Dear Frank:

I am inclosing receipt from Mr. Olsen for his salary check and expense check.

Mr. R. Rogers should go on the pay roll commencing May 31, at a salary of $40 per week.

My letter of June 6th had an account of 40 some dollars for Mr. Olsen covering expenses of driving his Buick through from the coast. I did not know whether this was to be allowed or not, but you did not say anything about it in your letter of June 9th and I suggest that you write to Mr. Olsen direct at the Pennsylvania Hotel regarding this as I do not know how much longer I will be in Indianapolis as I expect to go to Michigan to work on truck sales as soon as the state highway department business is settled here in Indianapolis.

Very truly yours,

H.G. Engel
HGE/PEC

On the same date Olen replied to Engel's June 6th letter to Gause. Olen made no effort to moderate the sting in his words.

June 11, 1932.

Mr.H.G. Engel,
532 East 54th St.,
Indianapolis, Indiana.

My dear Mr. Engel:-

Replying to your letter with reference to hiring Mr. R. Rogers at $40.00 per week, this is entirely out of the question. You and Mr. Olsen were authorized to secure a helper to help Mr. Olsen if he had some special work - someone that could be hired a day at a time.

I cannot see how Olsen can keep busy when the car is down waiting for Miller to bring parts and there is no reason for him to have a helper.

If you have obligated yourself, pay him up and let him go and tell Olsen when he actually needs a helper he can be hired. I do not see how Olsen can be busy when he is waiting for parts.

If all the parts could advantageously be loaded into the truck and be brought here to Clintonville, I think it is the best thing you can do. Will you look this up and let me know at once.

Yours very truly,
FOUR WHEEL DRIVE AUTO CO.

President

W.A.Olen,
EL

Olen wired a painful decision to Engel.

June 13, 1932.

H G ENGEL
532 EAST 54TH STREET
INDIANAPOLIS INDIANA

FOLLOW INSTRUCTIONS OF MILLER ON RACING CAR AS TO BEARINGS AND CRANKSHAFT

Walter A Olen
OLEN

WAO:EL

Engel's account was reimbursed for cost of gear transaction.

June 13, 1932.

Mr. H.G. Engel,
532 E. 54th St.,
Indianapolis, Ind.

Dear Mr. Engel:-

Enclosed herewith is deposit slip showing an amount of $69.32 which has been deposited to your account. This covers the gears which were ordered by Mr. Miller from the Carl H. Wallerich Co.

Yours very truly
FOUR WHEEL DRIVE AUTO CO.

Secretary.

HW
Encl.

Account rendered for Operation Race Car to date:

June 13, 1932

Mr. Olen:

Following is an account of the publicity received in newspapers and trade papers on the four wheel drive racing car entry to date.

Total value of publicity received in newspapers and trade-papers - - $107,409.16

Number of Articles - - - - -699

Total circulation of these articles - - - - - - - 40,927,169

Number of States in which publicity appeared - - - - - 39

Publicity also appeared in Newfoundland and Ontario provinces.

The value of the free publicity was figured at the flat reading notice rate for each individual newspaper. It is rather difficult to arrive at an accurate estimate in dollars and cents of the value of this publicity, inasmuch as the four wheel drive racing car is sometimes mentioned in only part of an article and yet that article, as a whole, deals with racing cars. Other times the article is 100% FWD and sometimes appears on the front page in headlines. We have taken, however, the flat reading notice rate for all articles.

Very truly yours,

W. M. Hanson
Advertising Manager
PGS

Engel, unperturbed, but again omitting Olen's name from his telegram, replied with logic to Olen's stern directive of the 11th.

GU4 79 NL COLLECT=INDIANAPOLIS IND JUN 14 1932

FOUR WHEEL DRIVE SALES CO=
CLINTONVILLE WIS=

CAR COMPLETELY DISASSEMBLED STOP DO NOT BELIEVE IT ADVANTAGEOUS TO SHIP CLINTONVILLE AS IT WOULD HAVE TO BE ASSEMBLED IN PART AND THEN DISASSEMBLED TO PUT IN BEARINGS STOP ROGER RECOMMENDED BY MILLER AS HE DID NOT BELIEVE IT ADVISABLE TO HAVE MEN WORKING THAT WERE UNFAMILIAR WITH CAR STOP HE AND OLSON HAVE BEEN BUSY ALL THE TIME BUT ARE THRU TOMORROW STOP CAR ALL READY EXCEPT BEARINGS AND WILL TAKE ABOUT THREE DAYS AFTER THEY ARRIVE=

H G ENGEL.
809AJUN 15

Ira Vail, who also organized the Syracuse A.A.A. Championship 100-miler, tried to tempt the Four Wheel Drive Auto Company's entry.

GU 15 50 DL 3 EXTRA=SYRACUSE NY JUN 14 1932 1243P

W A OLEN,PRESIDENT=
FOUR WHEEL DRIVE CLINTONVILLE WIS=

WILL APPRECIATE EARLY TELEGRAPHIC REPLY WHETHER YOU WILL HAVE CAR HERE FOR JUNE TWENTY FIFTH RACE THIS INFORMATION DESIRABLE FOR PROPER PUBLICITY STOP IN EVENT YOU SEND CAR HERE FOR TRYOUTS SEVERAL DAYS IN ADVANCE AND FIND CAR NOT IN PROPER CONDITION YOU MAY WITHDRAW ENTRY THANKS=

IRA VAIL CARE AUTOMOBILE CLUB.

116P.

Olen snapped his reply by economical night letter.

June 14, 1932.

IRA VAIL
CARE AUTOMOBILE CLUB
SYRACUSE NEWYORK

CAR NOT IN SHAPE

OLEN

Engel's memo of June 15th to Frank Gause illustrates the great buying power of the dollar at that time: The quibbling over $6.10 and getting a five-main-bearing crankshaft ground for $18 are both impressive.

To Mr. Frank Gause at Clintonville, Wis. Date 6/15/32
From E.G. Engel at Indianapolis, Indiana

Dear Frank:

I am inclosing some bills pertaining to the race car.

There are three to the Merz Engineering Company for machine work done on the gears totaling $41.60. These have not been paid and check should be sent direct to the Merz Engineering Co.

There is a question on their invoice number 3211 for $6.10 and I believe that you should take this up with Mr. Miller when he comes to the factory as he promised to correct the oil trouble at no extra expense to us and this particular invoice pertains to the oil trouble.

I am also inclosing an itemized list of $26.40. These bills have been paid by Mr. Olsen and check should be sent to him for this amount.

In regards to the telephone calls. Olsen was instructed to make no changes except those authorized by Miller and it was necessary for him to get in touch with Mr. Miller.

As you know the parts on this car are very delicate and will not stand any banging around and Mr. Olsen believed it advicable to get a special made box with compartments to put these extra parts in as a part that would cost 40 or 50 dollars might easily be damaged if it was not properly handled.

It has been necessary for Mr. Olsen to run back and forth from the speed-way to town to pick up parts and supplies from the machine shops etc. and that explains his charge of 25 gallons of gasoline for the buick.

I am also inclosing a bill from the Eagle Machine Company for $18.00 for regrinding the crank shaft. It was paid by myself. I also paid Mr. Rogers for two weeks and two days salary at $40.00 a week or $93.34 making a total due me of $111.34 which I would appreciate your depositing to my account at the bank.

Mr. Rogers is no longer on our pay roll, but it will be necessary to hire a helper for a couple of days when the bearings arrive from the coast.

As far as I know now the only additional expenses that there will be will be a cost of $15 for balancing the crank shaft and line boring the bearings when they arrive.

Very truly yours,

H.G. Engel

H.G. Engel
HGE/PEC

Olson's services were regarded as "labor" in Clintonville. No one there seemed to realize that they were getting a master machinist and racing mechanic—an excellent practical engineer in fact for $50 a week.

June 15, 1932.

Mr. E. W. Olson
c/o Pennsylvania Hotel
Indianapolis, Indiana.

Dear Sir:

Enclosed herewith you will find check No. 2250, amounting to $50.00, covering labor for period ending June 11, 1932.

Kindly acknowledge receipt of the above check and oblige.

Yours very truly
FOUR WHEEL DRIVE AUTO CO.

Secretary

HW
Encl.
Date Received (June 17 32)
Signature (Ernest W. Olson)
(Thanks.)

Another patient, defensive status report from Engel to Olen:

To Mr. W.A. Olen at Clintonville, Wisconsin Date 6/15/32
From H.G. Engel at Indianapolis, Indiana

Dear Mr Olen:

Confirming my night letter Mr. Rogers completed his duties today and he was hired at the request and recommendations of Mr. Miller as Mr. Rogers worked for Mr. Miller in the manufacturing of this car and was familiar with its construction. Mr. Miller was of the opinion that we should have an experienced man help Olsen and he thought that the salary of $40 a week was very reasonable.

I have been out to the track five or six times at different times of the day and have found Mr. Olsen and Mr. Rogers busy and they did not know when or if I was coming out.

It was not our intention to keep Mr. Rogers on the pay roll when there was no work to be done. I am not familiar enough with mechanics to know if there were any shorter cuts to the work, but I think the boys were honest and were not trying to put anything over.

I just returned from the track and Olsen will begin to reassemble the axles tomorrow as they have just completed lapping in the new ring gear and pinion. He expects to complete this work by the time the bearings get here so at that time all he will have to do is to have the bearings lined bored in the case and to assemble the motor. He estimates this work will take about three days after the bearings arrive.

In order to have shipped the car to Clintonville it would have been necessary to practically re-assemble the car as it would not be practical to ship these parts loose or boxed as there was danger of something happening to the parts in transit and it would then require the disassembling after the car got to Clintonville to install the bearings and I was not sure if we had machinery to line-bore which would have meant sending it to Waukesia or Milwaukee to get this done.

The highway commission had expected to meet today to decide on trucks, but their meeting was postponed on account of the death of a relative of Mr. Wedeking, chairman and they are not sure in the department, but they think they will meet either Thursday or Friday. Mr. Boren, the equipment inspector and brother of Mr. John Boren, one of the commissioners assures us that we are still in the picture and feels we will be awarded some of the business. None of them have told us exactly how many to expect.

I sincerely hope that my actions have met with your approval.

Very truly yours,

H.G. Engel
HGE/PEC

Gus Schrader, a great dirt track driver, provided a nice straight-from-the-cockpit testimonial about the car:

Cedar Rapids, Iowa.
June 16, 1932

Mr. H. G. Engel,
c/o General Equipment and Supply Co.,
31 W. 11 Street,
Indianapolis, Indiana.

Dear Mr. Engel:

Spent a few minutes in Indianapolis Monday but was unable to get in touch with you. Understand the car is progressing a little slowly. Olson said he didn't know whether or not you would campaign it on the dirt. If you do, I would certainly like a chance to drive it. Undoubtedly, I could get some good publicity for you, as other years, I have been able to get some nice contracts with State Fairs.

Am enclosing herewith a testimonial on the Four Wheel Drive, which I am glad to send to you.

Yours very truly,

Gus J. Schrader,
P.O. Box 217

GJS:m
enc

Cedar Rapids, Iowa.
June 16, 1932.

The Four Wheel Drive Auto Co.,
Clintonville, Wisconsin.

Gentlemen:

For fifteen years, I have been driving race cars of all types, but since having driven the four wheel drive at Indianapolis this year, I sincerely believe their handling is incomparable. There is no slueing or skidding going in or coming out of the turns.

Too, the acceleration is so much smoother since the power is being transmitted to four wheels simultaneously, which is not done on any other type drive. The same is true of deceleration.

I firmly believe in the principle of the Four Wheel Drive.

Yours very truly,

Gus J. Schrader

Gus J. Schrader.

Engel to Olen, June 20th. The would-be observer was Colonel Art Herrington, of the Marmon-Herrington Corp. and Chief of the Speedway's Technical Committee. The implications are infinite.

To Mr. W.A. Olen at Clintonville, Wisconsin Date 6/20/32
From H.G. Engel at Indianapolis, Indiana

Dear Mr. Olen

I am attaching letter received from Mr. Schrader regarding the FWD race car. [see previous letter]

Mr. Harrington sent word to Mr. Olsen that he would appreciate it if Mr. Olsen would inform him when he got the car put together and was going to try it on the track as Harrington would like to watch its performance.

I told Mr. Olsen that I was not sure whether you would want him to go out of his way to have Mr. Harrington watch the performance of this job and I would appreciate your instructions on this matter.

Very truly yours,

H.G. Engel

H.G. Engel
HGE/PEC

Another paycheck for day laborer Olson. Olson consistently added his "thanks" to his signature acknowledging receipt.

June 21, 1932

Mr. E. W. Olson
c/o Pennsylvania Hotel
Indianapolis, Ind.

Dear Sir:

Enclosed herewith you will find check No. 2251, amounting to $50.00, covering labor for week ending June 18, 1932.

Kindly acknowledge receipt of the above check and oblige.

Yours very truly
FOUR WHEEL DRIVE AUTO CO.

Frank Gause
Secretary.
HW
Encl.

Hanson had the imaginary value of publicity received in printed media figured out to the last cent, along with the exact number of papers.

June 21, 1932

Mr. Olen:

Following are the final figures on the amount of publicity received in newspapers and trade papers on the four wheel drive racing car.

Total value of publicity received
in newspapers and trade papers - - - - - - - - - - - 131,192.61

Number of Articles - - - - - - - - - - 828

Total circulation of these articles - - - 48,049 ,827

Number of states in which publicity appeared - - - - - - - - - - 40

Publicity also appeared in Newfoundland and Ontario provinces.

Very truly yours

W.M. Hanson
Advertising Manager
PS

By June 27th, Engel apparently had returned to Clintonville. Olen would not reimburse Olson for what he had spent out of his own pocket on the car until he, Olen, had a showdown with Harry Miller.

June 27, 1932

Mr. Ernie Olson,
c/o Pennsylvania Hotel,
Indianapolis, Indiana.

Dear Ernie:-

I had a talk this morning with Mr. Olen regarding the racing car, and he desires that after the bearings arrive that you get it in first class shape and try it out enough to see that it is right, that is, if it is necessary to run ten, fifteen, twenty or twenty-five miles in your opinion to properly test it out, that is what he wants you to do.

He does not desire that you make an extensive run at this time, and just as soon as the car is ih good shape in your opinion, please wire him and he will give you instructions, as he has not determined as yet just what he is going to do with the car.

If possible, he would like Mr. Miller to see this job, and if you are in telephone conversation with him, would appreciate your asking him if he could come to Indianapolis and look the job over while it is running.

I talked with Mr. Gause, our Secretary, regarding your expense account, and he has temporarily held this up, the same as he has held up payment of some of the bills until Mr. Miller comes to Clintonville, as Mr. Olen desired to talk these matters over with him before payment is made.

I do not believe that I will be able to get back to Indianapolis for the test.

Very truly yours,
FOUR WHEEL DRIVE AUTO CO.,

District Sales Representative.

H.G. Engel
MCK

Hanson to Olson, June 27th; the Advertising Manager at work again:

June 27, 1932

Mr. Ernie Olsen
Hotel Pennsylvania
Indianapolis, Indiana

Dear Mr. Olsen:

We are wondering if you would take the time to write us a letter, telling us what you think of the FWD racing car.

Attached is a suggestion, but we would much rather you write us, telling us in your own words how you feel about this racer. Please do not sign the attached letter. We would rather have you write your letter on hotel stationery and send it to us.

We will appreciate whatever you do for us in this matter.

Very truly yours,
THE FOUR WHEEL DRIVE AUTO COMPANY

W. M. Hanson
Advertising Manager

PS

Attachment.

The Four Wheel Drive Auto Company
Clintonville, Wisconsin

Attention Mr, W. A. Olen

Dear Mr. Olen:

I have worked on many racing cars. This year was the first time I had the pleasure of working on a four-wheel-drive.

I think that the four-wheel-drive type of drive is not only practical for racing, but has many advantages over other types of construction. This car ought to make an exceptionally good record, for it can make better time in the turns with a wider margin of safety and, as a result, ought to make a higher average speed over a long distance.

I must say that the car is beautifully designed and finished throughout and that I was especially impressed with the simplicity with which the drive to all four wheels was accomplished.

Sincerely yours,

Olson's wages at least were paid punctually.

June 28, 1932.

Mr. E. W. Olson
c/o Pennsylvania Hotel
Indianapolis, Ind.

Dear Sir:

Enclosed herewith you will find check No. 2253, amounting to $50.00, covering labor for week ending June 25, 1932.

Kindly acknowledge receipt of the above check and oblige.

Yours very truly_
FOUR WHEEL DRIVE AUTO CO.

Frank Gause
Secretary.
HW
Encl.

Olson complied with Hanson's request in his own words, while using Hanson's last paragraph verbatim:

June 29, 1932

The Four Wheel Drive Auto Company
Clintonville, Wisconsin.

Attention: Mr. W. A. Olen

Dear Mr. Olen:

After being mechanic on all types of racing cars for the past sixteen years, I have had the pleasure of working on one of the Four Wheel Drive cars, which has certainly impressed me as the coming car.

I have never ridden in any car with the feeling of safety which I had after the first lap in the Four Wheel Drive job. The ease of taking the turns in this car at a higher rate of speed and with absolutely no skidding has sold me on the Four Wheel Drive principle in a far greater way than the four wheel brakes impressed me several years ago.

I must say that the car is beautifully designed and finished throughout and that I was especially impressed with the simplicity with which the drive to all four wheels was accomplished.

This car, within the next year, should break a great number of the present day track records, which will prove more conclusively that your FWD principle is correct.

Sincerely yours,

E W Olson.

Ernie Olson

This letter from Hal H. Timian dated June 28, 1932 is a cover letter and update of one dated June 1st, which was returned to the sender, unclaimed.

June 28th, 1932.

Messrs.
Four Wheel Drive Company,
CLINTONVILLE, Wisc.

Attn: Mr. Walter Olen.

Dear Mr. Olen:

Upon my return to Indianapolis after the race, we addressed the enclosed letter to you at the Lincoln Hotel which came back unclaimed.

As expressed in the previous letter, we wish to extend our facilities in helping to demonstrate the merits of Four Wheel Driving to the public.

Mr. McDonough called at this office last Friday and we were glad to learn of the rebuilding program and the correction of the engine lubricating problems experienced at Indianapolis.

If we can be of assistance, please advise directly of through Mr. Mcdonough or Mr. Olsen.

Yours very truly,
CONTINENTAL MOTORS CORPORATION

Hal. H. Timian

Hal. H. Timian
ENGINEERING DEPARTMENT.
HHT:EH

Timian's letter of the 1st was overflowing with kind and positive thoughts. To say that it merited at least a thank you is a deliberate understatement.

June 1st, 1932.

Mr. Olen,
Hotel Lincoln,
Indianapolis, Ind.

Dear Mr. Olen:

I wish to thank you for the kind invitation to the drivers' and mechanics' party you sponsored the night of the race. I regret very much not being there but I was so fatigued it was impossible to leave my bunk after a few hours rest.

Earnie Olsen informed me or your intention to enter your four wheel drive car in the Detroit races. In view of this, we sent the following wire:

"Please advise Bob and Earnie that Continental extends every facility in helping prepare Four Wheel Drive car for Detroit race stop Wire arrival time or how we can cooperate".

Our intention in doing this is to cooperate in helping to overcome the present motor problems that caused the Indianapolis failures.

The writer has been associated with Indianapolis and Dirt track racing for the past ten years. In all this time, we never have experienced anything like the "safety feeling" we enjoyed while riding at the high speeds around the track and turns with Bob McDonogh. We feel the added traction of four wheel driving greatly improves the accelerating and decelerating of race cars with a total absence of skidding or what is known as "sunfishing" before entering or leaving the turns.

In our estimation, your car will break all track and dirt records as soon as the engine will stay put as well as the rest of the driving units.

Again thanking you for your kind invitation and anticipating a more successful race car showing in Detroit, we remain,

Very truly yours,
CONTINENTAL MOTORS CORPORATION

Hal. H. Timian

Hal. H. Timian
ENGINEERING DEPARTMENT.
HHT:EH

But no. Its receipt was barely acknowledged by Olen, and Timian's letter of the 28th was simply ignored. Olen's note would seem to be an insulting rebuke. Perhaps it was a way of saying, "mind your own business"; perhaps Olen could not support the phrase, "the Indianapolis failure"; perhaps the reason for the rudeness was political: Olen bought his engines from Waukesha, who may have been in direct competition with Continental.

June 29, 1932.

Continental Motors Corporation,
Detroit, Michigan.

Gentlemen:- Attention - Mr. Hal T. Timian

Your letter of June 1st addressed to me at the Hotel Lincoln, Indianapolis, has just reached me.

The racing car has not yet been put into operating condition, but we expect that it will be on the track in a few days.

Yours very truly,
FOUR WHEEL DRIVE AUTO CO.
President.

Wallie A. Olen

W.A.Olen
ELV

It was probably Bill Sturm's idea to hire the Speedway and there to run five-hundred flat-out miles. This could result in a mass of new speed records for that distance and for all of those leading up to it. It was not until the 29th that Sturm was able to reply to Olen's question: "What would it cost?"

WILLIAM F. STURM

4010 N. New Jersey
Indianapolis

June 29, 1932.

Mr. W. A. Olen,
President,
The Four Wheel Drive Auto Company,
Clintonville, Wis.

Dear Mr. Olen:--

I was delayed in answering your letter of June 15 by illness.

The approximate cost of a 500 mile run at the Indianapolis motor speedway would be:

Rental and miscellaneous cost	$150.00
A. A. A. sanction	250.00
Bob McDonogh, one week salary	100.00
If he were successful in setting mark of 100 miles an hour, additional	200.00
If a new track record	250.00
His expenses, Detroit to Indpls. and return, period of one week, approx	100.00
A. A. supervision, Charles Merz and two others, approx. total	200.00
Timer'	50.00
My supervisor	250.00
Tires, etc.	300.00
Expenses of your own necessary men.	? ? ?
	$1,850.00

The weather is hot now and some other work of mine would prevent my helping you on this before the middle of September, which I consider a better time than right now.

I expect to leave for England on July 8, to bring back Kaye Don and Miss England III to race Gar Wood at Detroit September 3, 4, 5.

I do hope this will interest you in the fall, and that it will also give you definite information on the condition of the car for next year's race.

I might say that I would have the heartiest co-operation of Pop Myers in the run.

Sincerely,

W. F. Sturm

W. F. Sturm

Edward T. Phillips somehow received the impression that the H. Miller who spoke to him about racing the new car up Pikes Peak was the Four Wheel Drive Auto Company representative from Cheyenne, Wyoming.

July 1, 1932

Mr. R. Schmidt
Four Wheel Drive Auto Co.
Clintonville, Wisconsin

Dear Mr. Schmidt:

Your representative, Mr. H. Miller of Cheyenne, Wyo., confered with me today relative to driving your Four Wheel Drive race car in the Labor Day Auto Classic on Pikes Peak.

At present I have nothing definite in view and would gladly consider any offer you might have in mind.

In the Free for All Event of 1929 I won first place with a Stutz Special. In 1930 I won the Stock Car Event with a Cord front wheel drive. I am convinced I could break the standing record in the Free for All Event and unquestionably capture the 1932 title, provided I drove a car so designed and with sufficient power to do so. However, in this altitude it would require a change of the compression ratio, also gear ratio, which I am fully equipped to do and render any change as would be necessary. With your assistance, I am positive we could make the alterations to meet the condition here.

I understand this car carries a Milller motor and any information pertaining to same and the type of construction throughout the car that you can supply me with, would be greatly appreciated.

Hoping to receive a favorable reply from you at an early date regarding these matters, I am

Yours very truly,

Edward T. Phillips

Edward T. Phillips
ETP/MM

A handwritten letter from Ernie Olson to Bud Engel [is this H.G. Engel?] reads:

July 1 – 32

Dear Bud.

Expect to have the car ready some time Tues July 5 and will give it a run for about 20 or thirty laps. You know it will have to be run slow to start with. Really got a nice job on the bearings and will inspect the rods before leaving here.

Promoted Stein for a set of 18-650 experimental tires. (No charge) would not like to make many fast laps on them.

Harry Miller phoned me Thurs morning but I missed him. He told the Mrs. he would see me soon. I expect him to call me from Syracuse Sat nite, and will ask him to try to be here next week.

There should be plenty of hotshot drivers in next week. Will get some one then to drive the car for me. it would hardly be worth while for Mac to come down to drive it for such a short run.

Will wire you Tues or Wed. how we are gettin along.

Yours Truly,

Ernie Olson.

The usual receipt form from Gause to Olson to Gause.

July 5, 1932.
Mr. E. W. Olson
c/o Pennsylvania Hotel
Indianapolis, Ind.

Dear Sir:

Enclosed herewith you will find check No. 2323, amounting to $50.00, covering labor for week ending July 2, 1932.

Kindly acknowledge receipt of the above check and oblige.

Yours very truly
FOUR WHEEL DRIVE AUTO CO.

Frank Gause
Secretary.
HW
Encl.
Date Received________
Signature____________

Don't call us. We'll call you.

July 6, 1932

Mr. Edw. T. Phillips
308 South Tejon St.
Colorado Springs, Colorado

Dear Mr. Phillips:

This acknowledges your letter of July 1.

Whereas we appreciate your interest in our four-wheel-drive racing car, I am afraid that we are not in position to make you any promises for the Labor Day Races.

Thus far we have not entered the car in any additional races and no program has been laid out for same. Should we require a driver for the Pikes Peak Race we shall keep you in mind.

Thanking you for your interest, we are

Yours very truly,
FOUR WHEEL DRIVE SALES CO.

R.H.Schmidt,
General Sales Manager

VIB

Olson to Engel, in a July 6th telegram: "CAR RUNNING GOOD EXPECT MILLER AND SALSRAUGH HERE TOMORROW WILL WIRE RESULTS OF TEST."

Telegram, Miller to Olen, July 11th; exciting news, eternal haggling:

GU16 91 DL=AU INDIANAPOLIS IND JUL 11 1932 323P

W A OLEN=
F W D AUTO CO CLINTONVILLE WIS=
LOUIE MYERS DROVE THE CAR TODAY AND WAS VERY MUCH ENTHUSED MADE LAPS IN ONE SEVENTEEN AND TWO FIFTHS HAD EXPERIMENTAL TIRES AND TREAD CAME OFF BUT THE CAR NEVER SKIDDED THE CAR IS IN GOOD SHAPE SO PLEASE ADVISE ERNIE OLSEN WHAT TO DO MY UNDERSTANDING WAS THAT YOU WOULD PAY HIM HIS GASOLINE EXPENSES TO AND FROM THE COAST WOULD LIKE TO HAVE YOU ALSO PAY THE BILLS HE HAS INCURRED HERE I WILL ADVISE YOU LATER WHEN I WILL LEAVE FOR THE COAST AND WILL LIKE TO SEE YOU=
HARRY A MILLER.
35 6P.

Olson wires Engel, July 12th; Olen did not reply to yesterday's telegraphed question: "WIRED MR OLEN YESTERDAY THAT CAR IS READY FOR SHIPMENT WHAT DO YOU WANT ME TO DO WITH CAR"

Olen to Olson, July 12. So long to the Speedway:

July 12, 1932

ERNIE OLSON
PENNSYLVANIA HOTEL
INDIANAPOLIS INDIANA

LOAD CAR IN TRUCK AND BRING HERE

OLEN

Another week of precision "labor" and high responsibility, another fifty dollars. 20,000 weeks, a million dollars.

July 12, 1932

Mr. E. W. Olson,
c/o Pennsylvania Hotel,
Indianapolis, Indiana

Dear Sir:-

Enclosed herewith you will find check No. 2324, amounting to $50.00, covering labor for week ending July 9, 1932.

Kindly acknowledge receipt of the above check and oblige.

Yours very truly,
FOUR WHEEL DRIVE AUTO CO.

Frank Gause
Secretary.
OM

Encl.
Date Received__________
Signature______________

Pop Myers to the Four Wheel Drive Auto Company's Chief Engineer, always gracious, always professional:

July 13, 1932.

Mr. H.B. Dodge,
% Four Wheel Drive Auto Co.,
Clintonville, Wisconsin.

Dear Sir:

Some of the specifications which have governed our recent 500-mile races seem to be under considerable discussion by a number of prominent engineers about the country.

Before formulating rules for the 1933 500-mile race we would appreciate very much having the benefit of the experiences and opinions of as many of these engineers as possible.

With this thought in mind we are arranging a meeting to be held in Detroit Monday morning, July 25th, at eleven o'clock at the Book-Cadillac Hotel, at which time all, or any part of the day necessary, will be devoted to discussions.

It is our hope that you will be able to attend this meeting. If it is not possible for you to be present personally we shall appreciate your sending an able representative to take your place.

You are invited to be our guest at luncheon, and Captain Rickenbacker and the undersigned are looking forward with pleasure to seeing you on the above named date.

Very truly yours,
INDIANAPOLIS MOTOR SPEEDWAY CORPORATION.

T.E. Myers

T.E. Myers
Vice-President.
tem-ed

The Indian wrestling over bills continues in this letter from Engel to Miller:

July 18, 1932

Harry A. Miller, Inc.
6233 Gramercy Place
Los Angeles, California

Dear Mr. Miller:-

We are inclosing our check for $174.55 to cover your invoice No. 23083 dated June 3 for $137.00 covering two connecting rods, one intake valve and one exhaust valve and $37.55 covering the amount due you for money expended at the Indianapolis Race in our interest.

It is my understanding from Olson that you accepted the return of the wheels so we deducted $84.00 from your statement submitted at that time for $121.55.

We are returning to you your invoice No. 23099 for $175.00 covering the cost of rebabbitting and machining eight connecting rods and one set of main bearings, as it was my understanding with you that this was to be done at no charge to us. You also agreed to fix the oil trouble and there are two bills that Olson turned in to us one for $2.00 and the other for $3.50 that we feel should be paid by you and we are inclosing them.

Yours very truly
FOUR WHEEL DRIVE SALES CO.

Sales Department

H.G. Engel
MIM

Inclosures -

Bryan Saulpaugh had come up with a very attractive package for the Four Wheel Drive Auto Company, for whose eyes this letter had been written. It led to nothing.

(COPY)

HANKINSON SPEEDWAYS

Ralph A. Hankinson, Gen. Mgr.

New York Office
1560 Broadway

Reading, Pa. Office
151 N. Fifth Street

July 18, 1932

Mr. Bryan Saulpaugh,
1631 21st Avenue,
Rock Islands, Ill.

My dear Bryan:

With reference to our discussion of the possibility of you obtaining a four-wheel-drive car for exhibition purposes in the East, I beg to advise that in the event you do obtain this car it will be possible for us, through the cooperation of the Fairs at which you will race, to obtain suitable exhibition space for the car at each or these Fairs without cost.

I believe the exhibition of this car, suitably arranged and accompanied by explanatory sign cards, and possibly including your services at the exhibition for a period each day during which you could answer inquiries, would prove to be a fine bit of exploitation for the owners. Further, it could be arranged for you to drive the car for exhibition purposes on each afternoon on which we will race, to be accompanied by suitable announcements over the public address system, which as you know are directly under our control on racing days.

Further, we would be pleased to cooperate with you in obtaining valuable publicity in newspapers in, and near, the cities where you will race for the car. You, of course, are familiar with our valuable publicity contacts here in the East and have some idea of just what we could do in this connection.

If you are successful in your efforts to make a connection with the owners and sponsors of this car, it would be well for you to advise me at the earliest possible time so that we could effect necessary arrangements with the individual Fair managements.

With kind regards and best wishes, I am

Cordially yours,
(signed) Ralph A. Hankinson

RAH*Y

Olen to Vail, whose letter of July 13th is not in the file. Now the car was for sale. The "new sub-transmission" no doubt was a new transfer case, as the Miller people referred to the unit.

July 19, 1932.

Mr. Ira Vail,
80 Bellows Lane,
Manhasset, New York.

My dear Mr. Vail:-

I have your letter of July 13th with reference to the racing car.

It has just arrived at our plant and is now on our show room floor. I think for the time being, at least, that we shall keep this for exhibition purposes.

Mr. Harry Miller is to stop here on his way back to California. He has been unavoidably delayed much longer than any of us expected, but I would like to see him before making final decision.

If you know of anyone who wants to buy the car, we would be willing to sell it. We would be willing to make some allowance for advertising that we would get out of it. The car is now in very fine condition. It made some twenty laps on the Indianapolis Speedway last week and was up to a speed of 117 miles per hour. There are three, if not four, different sets of gear ratios with the car. A new sub-transmission is being installed which will further improve the oiling facilities of the car.

Yours very truly,
FOUR WHEEL DRIVE AUTO CO.
President

W.A.Olen,
ELV

Vail replied by return mail, with an unsolicited testimonial and a proposition for his own acquisition of the car.

July 21, 1932

Mr. Walter A. Olen,
The Four Wheel Drive Auto Co.,
Clintonville, Wis.

My dear Mr. Olen:

I have received your letter of July 19th, with references to your wanting to sell your racing car.

Please advise me what price you are asking for the car and also what terms you will accept? I would like to buy this car providing the price is right, as I feel certain that it can win where ever it is raced under the proper management.

At the present writing I do not have any available cash but if you will let me take the car and race it I will agree to pay for it out of its earnings plus so much a race at the different fairs that I am conducting. In other words I will assure you of the full selling price of the car, with the car itself up for collateral until it is paid for.

If you will investigate the references I sent you some months ago, I am sure you will find that I always meet my obligations. As I wrote you I had in mind building a Four Wheel Drive Car several years ago and I am thoroughly sold on that principle. It is only a question of time when all of the racing cars will be four wheel drives.

I understand that Detroit will have a championship race on September 4th, and there is a good possibility of Syracuse holding a race in October, so I can take in these races together with about ten others here in the east before racing the car in California this fall.

Trusting that you will see your way clear to let me have the car on the above mentioned terms, I am

Very truly yours,
(signed) Ira Vail

Olen to Saulpaugh, unmoved.

July 25, 1932

Mr. Bryan Saulpaugh,
1631 - 21st Avenue,
Rock Island, Illinois.

Dear Sir:-

I am enclosing and returning the letter from the Hankinson Speedways. I will advise you later as to our disposition in this matter.

Very truly yours,
FOUR WHEEL DRIVE AUTO CO.

W.A. Olen

President

MCK
Encl.

Olen replies in "My dear" mode, to Vail: Speak to me of cash.

July 26, 1932.

Mr. Ira Vail,
80 Bellows Lane
Manhasset, New York.

My dear Mr. Vail:

Replying to you letter of July 21st, the price of the racing car, complete with racing tires, is $16,000.00. On a cash basis, we would allow something for advertising.

The car has just recently arrived at our plant and now seems to be in very good condition. There are a good many inquiries about the car and several people have been here to look it over.

While we would not expect all cash, we would require a substantial down payment in case of a sale.

Yours very truly,
FOUR WHEEL DRIVE AUTO CO.
President.

W.A.Olen
ELV

In his stumbling prose, Hanson solicited another few testimonials:

July 29, 1932

Mr. Barney Oldfield
c/o Indianapolis Speedway
Indianapolis, Indiana

Dear Mr. Oldfield:-

We would like very much to have you write us,telling us what you think of the FWD racing car. If you can find it convenient to do so, we certainly would appreciate very much your writing us a letter, telling us in your own words how you feel about this car.

Attached are letters from Robert McDonogh and Ernie Olsen, in which they express themselves regarding the racer.

I want you to know that I appreciated the cooperation you extended me and our organization during Race Week at Indianapolis. We were pleased, indeed, with the way you handled things and we want you to know that it certainly was a pleasure to work with you.

Very truly yours,
THE FOUR WHEEL DRIVE AUTO COMPANY

W. M. Hanson
Advertising Manager

PS

The next day, July 30, 1932, Hanson sent a slight variation of the same letter to Mr. Louis Meyer c/o the Indianapolis Speedway, and to Mr. Phil Schaefer, also c/o the Indianapolis Speedway.

Chapter 8

1932—The Brisko Campaign

Illustration 8-1 Walter Olen, president of the company, helmeted and ready to charge with home state driver Frank Brisko. (GBC ex FWDCO)

How it was that Walter Olen learned that a young racing driver and practical engineer from Milwaukee, Wisconsin, wanted to get his hands on the four-wheel-drive car we do not know. It is only 125 miles from Milwaukee to Clintonville, and it may well have been that Frank Brisko was one of those who, as Olen had just told Vail, had come to look the car over. While Olen had dismissed Saulpaugh brusquely, he saw possibilities in Brisko and personally launched a character check on the man.

August 13, 1932.

Indianapolis Motor Speedway Corporation,
Indianapolis, Indiana.

Gentlemen:-

We would appreciate any information you could give us on Frank Brisko. We are particularly interested in honesty, truthfulness and general conduct among other drivers.

Any information you give us will be treated confidential and very much appreciated.

Yours very truly,

FOUR WHEEL DRIVE AUTO CO.

President.

W.A.Olen,
ELV

Olen had his Milwaukee Branch Manager run a check. The results were so negative that it is surprising that his curiosity about Brisko did not end there.

August 13, 1932.

Mr. W.A.Olen,
Four Wheel Drive Auto Co.,
Clintonville, Wis.

My Dear Mr. Olen:

In the past few days I have made several contacts, in regards to checking up on Mr. Frank Brisko, of this city.

Mr. Thomas Marchese, race car owner and promotor, Mr. Grover Horn, one time racer and builder, Mr. Jules Burbach, and Mr. Dooley, of the West Allis State Bank, are the Gentlemen that I have talked to.

Mr. Brisko has spent most of his time driving and building race cars.

It appears that he is not very steady, with any one owner, but that he is always jumping around from one car owner to another and trying to get someone to finance him. At present, a Mr. Atkinson of St. Paul, a department store owner, is financing Mr. Brisko and also ownes the car, a Miller, which he now has in his garage, at 60th and Greenfield Avenues.

Mr. Horn, who has driven race cars, and owns two of them now, says that all race drivers go strong for a time and then begin to lose their nerve and fade out of the picture. He believes, that as far as a winner is concerned, that Mr. Brisko is all through now. Mr. Brisko told Mr. Horn, personally, that he intended to try the Indianapolis track just once more and that if he did not hit then that he intended to quit driving for good.

Another thing, Mr. Brisko seems to have a tendency of talking, or bragging, too much among the drivers and they, the drivers, are all sour on him, for this reason.

Mr. Dooley, of the West Allis State Bank, said that Mr. Brisko used to have a small savings account with them, but that he has entirely withdrawn it.

This is about all the information I have, Mr. Olen, and I hope it is in line with what you desired.

Very truly yours,
FOUR WHEEL DRIVE SALES CO.

C. M. Bovee

C.M.Bovee.
Milwaukee Branch Manager.

However, Olen did give Brisko a chance to defend himself.

August 15, 1932.

Mr. Frank Brisko,
3124 N. 39th Street,
Milwaukee, Wisconsin.

My dear Mr. Brisko:

We have checked up on some of the statements that you gave us as to your earnings and as to your present arrangements and they do not seem to be corroborated.

Before committing yourself or making any announcement concerning our car, we shall want more information. We would like a statement of just what property you own and what it consists of and where it is located. We would also like an itemized statement of your winnings in the racing game last year.

We would also like a statement of the amount you are owing and to whom you are owing it. You must realize that we are entrusting you with a piece of machinery that is valued about $20,000, and it is necessary for us to have very definite information.

Yours very truly,
FOUR WHEEL DRIVE AUTO CO.

Walter A. Olen
President

W.A.Olen,
ELV

That was on August 15, 1932, and on that same day Pop [T.E.] Myers replied from the Speedway, painting a picture of Brisko that verged on being laudatory.

August 15, 1932.

Mr. W.A. Olen,
% Four Wheel Drive Auto Co.,
Clintonville, Wis.

Dear Mr. Olen,

Frank Brisko has been an entrant and a competitor in our 500-mile race for quite some years and it has been our experience with Mr. Brisko that he has always shown a disposition to be quite fair in his dealings and negotiations, has fulfilled his promises to us in every way and while we have no knowledge of his financial position, he has always paid his obligations to this Corporation as per agreement.

His conduct at this track has always been of the highest quality. He has been pleasant to have around and appears to be well liked by his fellow drivers. It is our opinion that his word may be depended upon and we have every reason to believe that he is honest. In addition to this he is looked upon as a careful driver and mechanic.

This is in answer to your letter of August 13th and the writer wishes to say that he is very happy to accommodate you in this matter and if there is any additional information that you might need, please do not hesitate to call upon us.

With all good wishes and best regards, we are

Very truly yours,

INDIANAPOLIS SPEEDWAY CORPORATION.

T.E. Myers
Vice President.
TEM:hr

On that same day, it occurred to Barney Oldfield to send a postcard to Olen from New York City. This is its message side, with another Oldfield testimonial, while the other side obviously carried a photo of "The Master Driver of the World" with his new Plymouth. Note the 1 1/2-cent stamp.

Perhaps Barney hadn't given the company back any of the money that he owed it, but that was all right. Olen liked the man to the extent that a photo of him smiling around his cabbage cigar made Olen's day. And he promptly wrote his friend to tell him so.

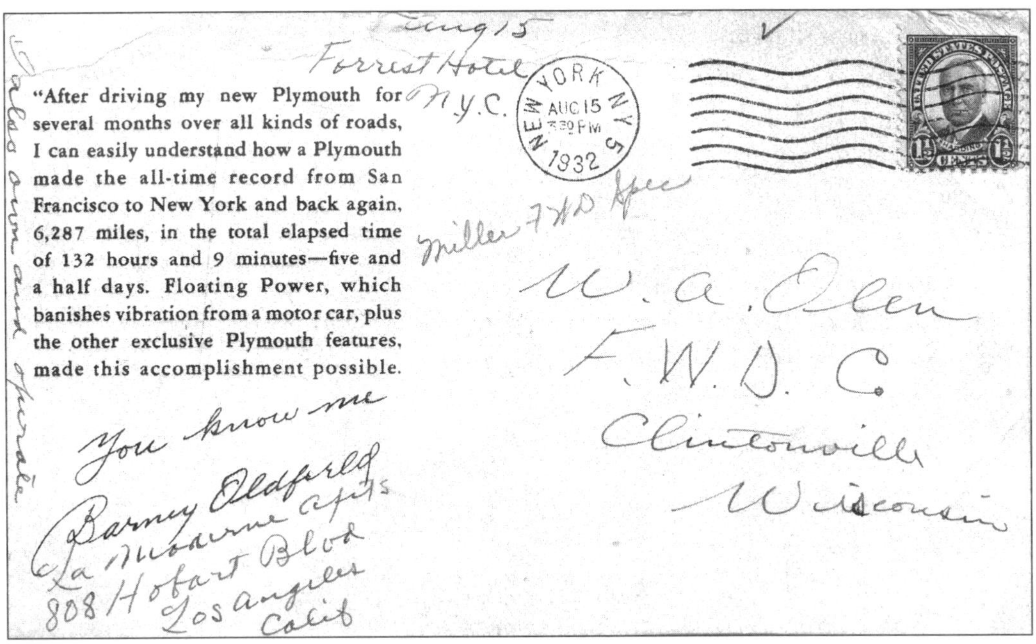

Illustration 8-2 Barney Oldfield's postcard to Walter Olen.

August 17, 1932.

Mr. Barney Oldfield,
Forrest Hotel,
New York, New York.

My dear Barney:

It rained yesterday - I came in from the Lakes this morning and found a bunch of irritating subjects on my desk for consideration and at the bottom of the pile I looked down and saw that magnificent smile, and that most beautiful cigar, of Barney Oldfield, and it just put new life and pep in me.

I want to thank you for sending me that postal card.

Yours very truly,
FOUR WHEEL DRIVE AUTO CO.

President

W.A.Olen,
ELV

The thought of that great showman may have influenced Olen's mood when he dashed off this simple message to a younger man who, for Olen, may have had some of the same right stuff:

August 17, 1932.

FRANK BRISKO
3124 NORTH 39th STREET
MILWAUKEE WIS

CAR AVAILABLE WHEN YOU NEED IT

OLEN

WAO:EL

Brisko wired back his appreciation that evening. Time would prove that Olen's confidence had been extremely well placed.

GU2 41 NL=MILWAUKEE WIS AUG 17 1932

FOUR WHEEL DRIVE AUTO CO=
ATTN W A OLEN CLINTONVILLE WIS=

THANKS FOR THE CONFIDENCE YOU HAVE PLACED IN ME WILL DO EVERYTHING IN MY POWER TO BE WORTHY OF IT WILL BE DOWN MIDDLE OF NEXT WEEK FOR CAR WILL TRY AND ARRANGE TO GIVE YOU A DEMONSTRATION AUGUST 28=

FRANK BRISKO.
AUG 18 806A.

Oldfield received Olen's note on the 19th and promptly wrote [in longhand]:

Aug 19 -32

W. A. Olen
Clintonville, Wis.

Dear Mr. Olen:

Thank you very much for your letter of Aug. 17th.

I was very proud to have managed your four wheel drive racing car in the last Indianapolis race, it is my idea of a racing car and if I were to build one or have one built it would certainly be a four wheel drive, you know. I drove your car in a practice spin and it just handled wonderfully. I am indeed sorry that car was not in Indianapolis in plenty of advance of race as I am sure that if we all had of had more time why the car would have made a much more creditable showing.

Wishing you all the success in the world.

You know me

Barney Oldfield

Barney Oldfield

As for Brisko, grand decisions had been made with remarkable rapidity. The Company would give the young man a try, leasing the car to him for ninety days. Here is the contract between them.

THIS INDENTURE made and entered into the 29th day of August, 1932, between the Four Wheel Drive Sales Company of Clintonville, Wisconsin, hereinafter called "Lessor" and Frank Brisko of Milwaukee, Wisconsin, hereinafter called "Lessee"

WITNESSETH:

That the Lessor hereby leases to Lessee and Lessee hereby agrees to rent from the Lessor, one racing car described and known as the "Miller Four Wheel Drive Special" now at Clintonville, Wisconsin, for a period of ninety (90) days and ending on November 29th, 1932.

It is expressly agreed and understood that said racing car may be driven by Lessee in races authorized and sanctioned by the American Automobile Association, and may be exhibited at fairs and shows in such places as the Lessee may select.

The Lessee is to furnish all repair parts,oil, gasoline, tires and all other repairs and expenses of any name and nature to keep said car in first class condition, all at his own expense, as may be required during the term of this lease and to put and keep said car in condition at all times,and to keep it in a clean and beautiful appearing condition,and to return it at the termination of this lease in as good condition as when received; usual and ordinary wear only excepted.

IN CONSIDERATION WHEREOF, the Lessee binds himself to pay as rent for said car a sum equal to twenty-five (25) per cent of the gross prize winnings or earnings whether earned through races or for exhibit or any other purpose that said car may earn as a result of the Lessee participation in races, shows, fairs or other exhibition or use of said car, said sum to be due and payable to Lessor at the office of the Four Wheel Drive Sales Company immediately after the participation in each event and upon payment of such prize money or earnings.

It is further agreed that Lessor shall not be liable to said Lessee for damages of any kind or from any cause whatever to the persons or property of Lessee, or any person claiming by, through or under him, and Lessee agrees to defend at his own expense any damage suits upon claims relating to said cause or the use thereof by the Lessee.

The Lessee further agrees that he would cause said car to be insured against fire at his own expense with a loss payable clause payable to the Lessor, and further agrees to pay all garage storage, transportation and other expense which may be involved in moving said car from one place to another; it being expressly agreed that the Lessor is not to be subjected to any expense of any kind or nature whatever during the term of this lease. If any parts in the car are broken or damaged in any manner, the Lessee is to provide such parts at his own expense and to keep said car in constant operating condition.

The Lessor reserves the right to inspect said car at any time and in case the Lessee fails to fully and completely live up to all the terms of this lease, the Lessor is privileged to take possession of same and to return it to its factory at Clintonville, Wisconsin without any liability to the Lessor whatever, in such case charging the lessee with any cost which may be incurred in transporting said car to its factory at Clintonville.

IN WITNESS WHEREOF, the parties hereto have hereunto set their hands and seals the day and year first above written.

Four Wheel Drive Sales Co., W.A. Olen, President
LESSOR

Frank Brisko
LESSEE

Attest:

Frank Gause
Secretary

It had been foreseen and agreed that Brisko would enter and drive the car in the upcoming A.A.A. Championship race at Detroit. He was therefore provided with a letter of introduction to the Company's dealer there.

August 29, 1932

Mr. Frank Brisko,

Attached is a letter of introduction to Mr. George L. Earle of the Earle Equipment Company, 6331 Tireman Avenue, Detroit, our dealer at Detroit. We are sure that they will be glad to meet you while at Detroit.

Very truly yours,
FOUR WHEEL DRIVE SALES CO.

S. H. Sanford,
Assistant Sales Manager.

MCK

The detailed bill of sale/lease form, made up the following day, lists the parts and materials entrusted to Brisko with the car, down to a sheet of gasket paper. And guess what we discover here? This is Miller four-wheel-drive chassis No. 2! It can be assumed that the so-called duplicate car, which always has been regarded as No. 2, was in reality No. 1, the prototype. It was the one on which the cut-and-try solutions were applied, as Harry Miller and his staff learned how to build such a chassis. The Four Wheel Drive Auto Company car was the true No. 2 car, the hopefully perfected "production" version. The car that Miller kept, then, corresponds exactly to the old outboard-brake front-wheel drive. This one was not the answer and Harry Miller was able to unload it on the Packard Motor Car Company for the full list price of the perfected inboard-brake model. We shall return to this subject later, in a surprising manner.

Brisko to Hanson, September 5th: The 100-mile Championship race held on Detroit's one-mile dirt track was called on the 84th lap, because of rain. Brisko's message touched only the immediately essential facts.

GU5 8=MILWAUKEE WIS SEP 5 300P

W M HANSON=
73 NORTH CLINTON AVE CLINTONVILLE WIS=

RAINED OUT RACE SATURDAY CAR AT DETROIT DEALERS=

FRANK BRISKO.
521P.

About a week later, Preston Tucker surfaced, with a night letter to Olen that is fairly packed with information. Tucker had not been asked to render this service; it was a pretext for renewing contact.

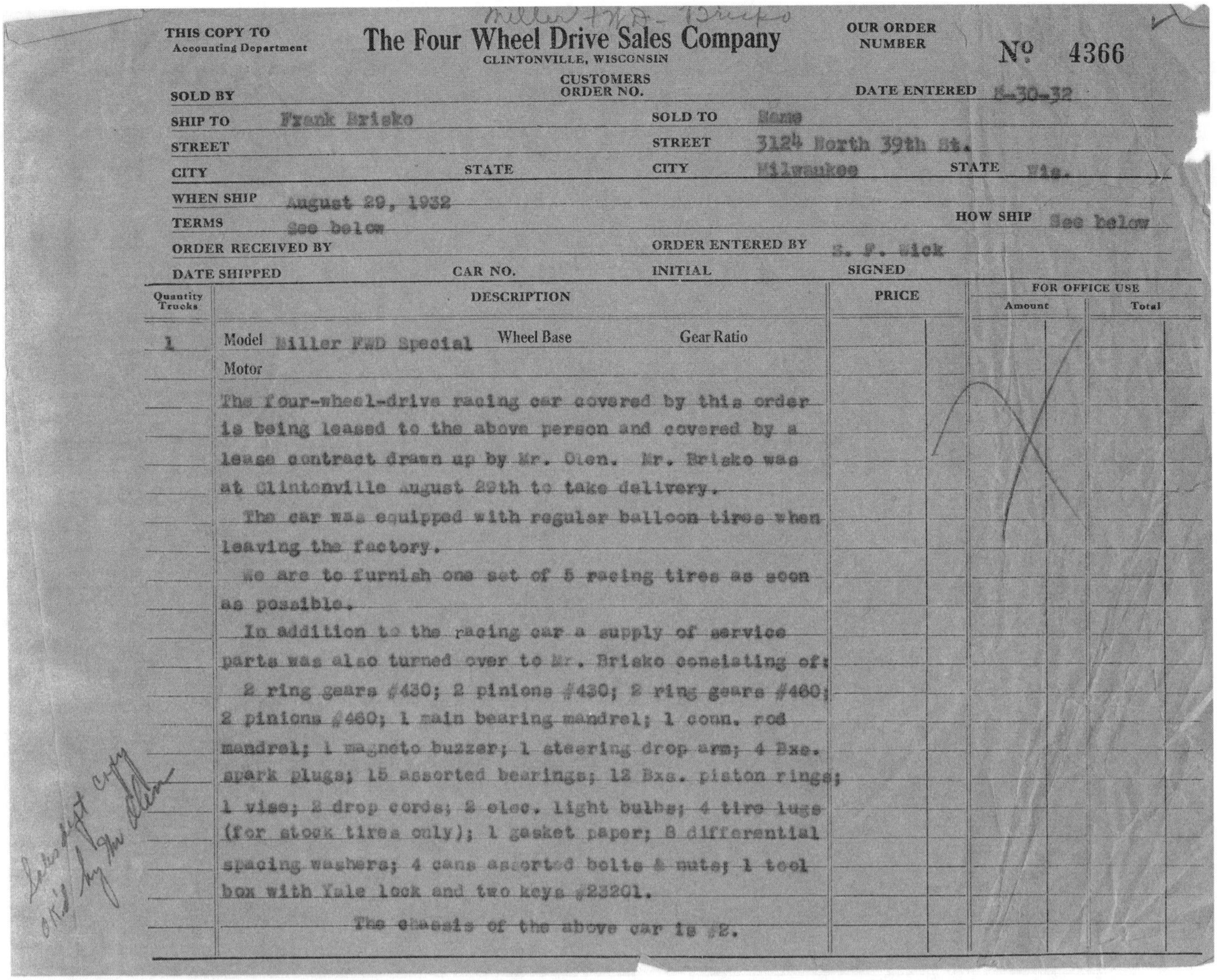

THIS COPY TO Accounting Department

The Four Wheel Drive Sales Company
CLINTONVILLE, WISCONSIN

OUR ORDER NUMBER Nº 4366

SOLD BY
CUSTOMERS ORDER NO.
DATE ENTERED 8-30-32

SHIP TO Frank Brisko
SOLD TO Same
STREET
STREET 3124 North 39th St.
CITY
STATE
CITY Milwaukee
STATE Wis.
WHEN SHIP August 29, 1932
TERMS See below
HOW SHIP See below
ORDER RECEIVED BY
ORDER ENTERED BY S. F. Wick
DATE SHIPPED
CAR NO.
INITIAL
SIGNED

Quantity Trucks	DESCRIPTION	PRICE	FOR OFFICE USE Amount	Total
1	Model Miller FWD Special Wheel Base Gear Ratio			
	Motor			

The four-wheel-drive racing car covered by this order is being leased to the above person and covered by a lease contract drawn up by Mr. Olen. Mr. Brisko was at Clintonville August 29th to take delivery.

The car was equipped with regular balloon tires when leaving the factory.

We are to furnish one set of 5 racing tires as soon as possible.

In addition to the racing car a supply of service parts was also turned over to Mr. Brisko consisting of:

2 ring gears #430; 2 pinions #430; 2 ring gears #460; 2 pinions #460; 1 main bearing mandrel; 1 conn. rod mandrel; 1 magneto buzzer; 1 steering drop arm; 4 Bxs. spark plugs; 15 assorted bearings; 12 Bxs. piston rings; 1 vise; 2 drop cords; 2 elec. light bulbs; 4 tire lugs (for stock tires only); 1 gasket paper; 8 differential spacing washers; 4 cans assorted bolts & nuts; 1 tool box with Yale lock and two keys #23201.

The chassis of the above car is #2.

Illustration 8-3 The lease form and list of parts entrusted to Frank Brisko.

GU2 50 NL=DETROIT MICH SEP 11 1932

WALTER OLEN,PRESIDENT FOUR WHEEL DRIVE AUTO CO=
CLINTONVILLE WIS=

BRISKO DROVE CONSERVATIVE SATURDAY FAIR SHOWING PICKED UP NAIL IN SECOND LAP RIGHT FRONT DRIVE JOINT FAILURE SIXTY EIGHT LAP WAS IN SIXTH POSITION MOVING UP FORCED OUT SIXTY NINTH LAP USED POSITIVE LOCK TOO MUCH POWER SPUN WHEELS NO DAMAGE PLEASE INFORM SCHMIDT AWAITING HIS INSTRUCTIONS REGARDS LETTER FOLLOWS=

PRESTON TUCKER.
808A.SEP 12.

Also on September 11th, Brisko wired three more crumbs of information to Hanson. Read "stop" after the word "joint."

```
GU1 10=YPSILANTI MICH SEP 11 542P
W M HANSON=
93 NORTH CLINTON AVE CLINTONVILLE WIS=
RUN 67 MILES BROKE UNIVERSAL JOINT ON WAY HOME=
FRANK BRISKO
459P.
```

On September 12th, Indianapolis Motor Speedway sent to known car owners a form letter and entry form for the 1933 Indianapolis 500.

September 12, 1932.

Four Wheel Drive Auto Company,
Clintonville, Wisconsin.

Gentlemen:

Entry blanks for the 21st Annual International Sweepstakes scheduled for Tuesday, May 30, 1933, are handed you herewith.

So that misunderstandings may be avoided, we urge you to carefully read the conditions under the heading "APPROVED SUPPLEMENTARY REGULATIONS"

--and, especially direct your attention to the paragraphs printed in red.

The entry list is now open and early nominations are solicited.

If you need additional blank forms, we shall be glad to supply them upon request.

Assuring you that we shall appreciate your support, we are

Very truly yours,

INDIANAPOLIS MOTOR SPEEDWAY CORPORATION.

T.E. Myers
Vice-President.
TEM-ED

Brisko also bent a wheel at Detroit. He promptly ordered a replacement from the manufacturer there, resulting in the following communication:

September 13th, 1932.

Four Wheel Drive Truck Company,
Clintonville, Wisconsin.

Gentlemen:

Mr. Frank Brisco, who drove one of your cars in Saturday's AAA Automobile Race at the Fair Ground here in Detroit, stated he was in need of a seventy spoke racing wheel for his car.

He advised us to ship this wheel to him in Milwaukee and apply it against your order H-04125 which is the order covering your tire equipment for the racing car from the Firestone Tire and Rubber Company.

However, it will be necessary for us to receive an order from you before we can ship this wheel to Mr. Brisco, therefore kindly either give us written authority to ship this wheel against the above mentioned order or send us another order covering.

Mr. Brisco states he wants this wheel shipped to him at 6014 National Avenue, Milwaukee, Wisconsin.

Very truly yours,

KELSEY-HAYES WHEEL CORPORATION

Dallas S. Kelsey

DSK:z

By September 14th, Miller, in Los Angeles, had received an order for and shipped to the Four Wheel Drive Auto Company two new U-joint sliding blocks, plus an inner and outer U-joint housing.

Having not heard from Miller, Brisko, taking no chances, sent this wire to Clintonville:

GU7 10=LN MILWAUKEE WIS SEP 15 1112A

MR BILLINGS=
CARE F W D AUTO CO CLINTONVILLE WIS=

DID NOT HEAR FROM MILLER WOULD ADVISE TO MAKE PARTS=

FRANK BRISKO.
1122A.

Compliance with this suggestion was set in motion the same day.

September 15, 1932

Mr. McConville
Production Dept:

Please make for FWD Miller Special racing car -

2 - Universal joint blocks
1 - Universal joint retainer, and
1 - Universal joint cover

as per sample joint left here by Frank Brisko.

These parts should be shipped as soon as possible to our Milwaukee Branch and addressed to Mr. Frank Brisko.

Yours very truly,

B. G. Donley
Chief Draftsman.
LAN

CC: Service Dept.
CC: Cost Dept.

Also on September 15th the tires, tube, and tire flap were received by Clintonville from the Milwaukee branch.

Olen replied to Preston Tucker, but there is no copy of that message in the file. Tucker replied on the 17th with valuable comments on the Detroit 100. He made known his activity as a hustler in Republican politics.

Detroit, Michigan.
9/17/32

Dear Mr. Olen:

Thought you people had forgotten me. Was a not until I received the most welcomed letter yesterday that I learned of Mr. Schmidt's operation. Certainly hope he is doing well and returns soon - he is a splendid man.

Believe me that FWD race car sure could get around this dirt track so darn much better it was hard to drive. I don't mean from a handling standpoint but purely a mental condition that was really hard to overcome. After about 40 laps he got confident and started to go would make one fast lap then two slow ones. The slow ones were about 1 second better than the leader was going. I say slow because when he stepped down it was so much faster and was doing it so easy that a good lap looked slow--on the second lap he picked up a nail - right rear - had a slow tire change lost 6 laps - Firestone loaned a spare wheel did not go on just right. On his 55th lap he started passing everbody inside of them or outside made no difference doing this only on the main stretch and south turn the back stretch was cut up the worst I have seen. Had rained hard and steady for four days before race making track soft - holes 10 inches deep that could not be missed several cars hit holes on north turn going thru fence. 4 cars forced out with bent axles and on the 67 or 68 lap he broke right front axle shaft, came out of socket - used center locking unit going around as rear drive this made quite a hit before he came in.

Had Frank started going in the beginning would of been no trouble have to lead race.

Harry Hartz said he is planning to convert the front drive that won the race this year into a FWD for next years 500 mile race.

Rose won driving Miller, Cummings and Miller. Chet Miller third Marr Special.

While waiting contract I have been vary active in the election. I managed a congressional candidate and at this writing we are three votes behind with three precincts to be heard from, very interesting, had a man who was strong at Washington to beat.

Yours truly,

Preston Tucker

P.S. I won delegate to Republican Convention.

The Four Wheel Drive Auto Company to Kelsey-Hayes: Normal caution.

Sept. 20,1932

Kelsey-Hayes Wheel Corporation,
Detroit, Michigan.

Attention - Mr. Dallas S. Kelsey.

Gentlemen:

We have your letter of September 13th and before mailing you our confirming purchase order for the wheel ordered by Mr. Frank Brisko, we ask that you give us a price on this wheel.

We trust to hear from you promptly.

Yours very truly,
FOUR WHEEL DRIVE AUTO CO.

Purchasing Department.

LJPinkowsky
ELV

On the 19th, Miller shipped to the Four Wheel Drive Auto Company the parts ordered on the 14th. It would seem that they arrived in Clintonville on the 22nd.

On the same date the parts were shipped back to Miller for the reason stated here.

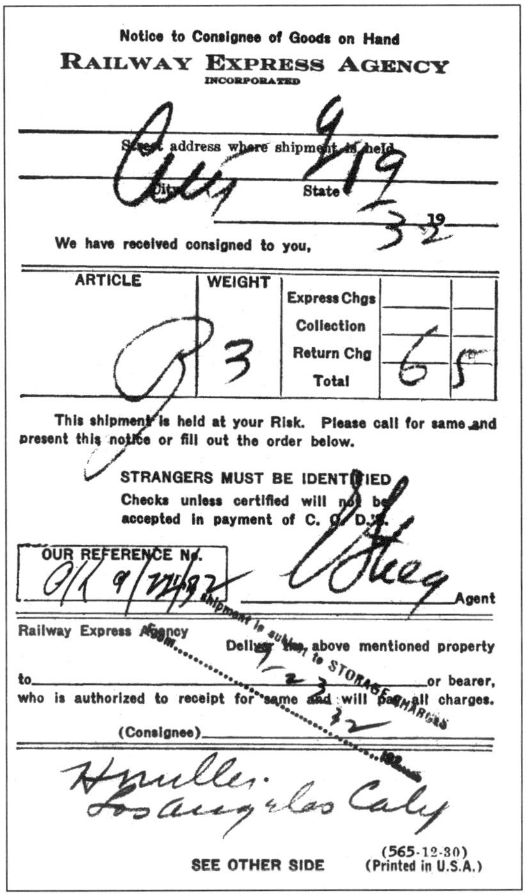

Notice to Consignee of Goods on Hand

RAILWAY EXPRESS AGENCY

INCORPORATED

Street address where shipment is held

City State

19 32

We have received consigned to you,

ARTICLE	WEIGHT			
	3	Express Chgs		
		Collection		
		Return Chg		
		Total	6	5

This shipment is held at your Risk. Please call for same and present this notice or fill out the order below.

STRANGERS MUST BE IDENTIFIED

Checks unless certified will not be accepted in payment of C. O. D.'s

OUR REFERENCE No.

9/22/32 Agent

Railway Express Agency

Deliver the above mentioned property

to ____________ or bearer,

who is authorized to receipt for same and will pay all charges.

(Consignee) ____________

Shipment is subject to STORAGE CHARGES

H. Muller. Los Angeles Calif

SEE OTHER SIDE (565-12-30) (Printed in U.S.A.)

Illustration 8-4 Railway Express Agency notice that parts were received.

9-22-32
Claim #5001
Report#5939

The parts on this claim were ordered by Frank Brisco for the racing car. The parts did not arrive soon enough and it was necessary to make up these parts here at the factory. Therefore, upon the arrival of these parts, they were not needed and have been instructed by Mr. Billings to return them to Mr. Miller. No credit will be allowed. claim closed.

On September 23rd, a message came from one of the Four Wheel Drive Auto Company's branch managers which testified to Brisko's sense of showmanship and business responsibility. It made the rounds of the Four Wheel Drive Auto Company top management.

9/23/32
To Branch Department at Factory
From J. R. Gray at Harrisburg Branch

Dear Sirs:

Mr. Brisko just stopped at noon with the FWD racer. He was very fine about showing the car to the State Highway officials, Mr. Schmidt, Kissinger, Fred Mason, all the mechanics at the State Garage, Mr. Keller of the Auditing Department, Mr. Meyers assistant chief engineer, and then over at the Bell Telephone Company Mr. Lutz and one other official whose name I did not catch, along with about one hundred and fifty curious people along the street. Mr. Lutz of the Bell Telephone Company is very interested and asked many questions, finally admitting one argument he has held out against us all this time, and acknowledging they do have speed and economy for their lighter work. He was very enthusiastic over the racer. The State Highway Engineers were very pleased that we stopped to show the car to them.

Mr. Brisko asked that this be called to Mr. Olen's attention as he wanted Mr. Olen to know that every place it was possible to stop and show the car in connection with our dealers or Branch that he is doing everything he can.

Yours very truly,

J.R.Gray
Branch Manager

On the 23rd, Dallas S. Kelsey, of Kelsey-Hayes Wheel Corporation, replied with information that should have been included in his original letter.

September 23rd, 1932.

Four Wheel Drive Auto Company,
Clintonville, Wisconsin.

Gentlemen: Attention--Mr. L.J. Pinkowsky

Have your letter of September 20th with reference to the racing wheel for Frank Brisko referred to in our letter of the 13th, and wish to advise that the standard price of our racing wheels is $21.00 each, including side ring.

Awaiting your further advice, we are

Very truly yours,

KELSEY-HAYES WHEEL CORPORATION

Dallas S. Kelsey
DSK;z

The Four Wheel Drive Auto Company to Miller: Take back your parts. This was pretty dirty pool, since (a) Miller had not been given a deadline, (b) the order to him had not been cancelled, and (c) he had filled it with very reasonable rapidity.

Sept. 23, 1932

Harry A. Miller, Inc.
6233 Gramercy Place
Los Angeles, Calif.

Gentlemen:

We have received an express shipment from you, consisting of:

2 - Universal Joint Blocks
1 - Universal Joint Inner Housing
1 - Universal Joint Outer Housing

These parts were ordered by Mr. Frank Brisco, for the Miller FWD Special Racing Car. However, this shipment was delayed too long and it was necessary that these parts be made up here at our factory. Therefore, when your shipment arrived, they were no longer needed and we are, therefore, returning them to you, under separate cover.

Upon the arrival of these parts, please cancel your invoice to us.

Yours very truly,
FOUR WHEEL DRIVE AUTO CO.

B. V. Walch
Service Department

BVW:I

The haste over the needed parts turned out to be due to the running, on September 24th, at Langhorne, Pa., of one of the few races left on the A.A.A. Championship Trail. One could risk life and machine against a miserly purse.

In spite of really severe leg burns due to overheating of the transfer case, Brisko told Hanson that the car performed perfectly.

GMB1 17 TDP PAOLI PENN SEPT 25 1932 1054A

W M HANSON=
CLINTON AVE FOUR WHEEL DRIVE CLINTONVILLE WIS=

FINISHED SECOND IN MAIN EVENT FIFTY MILES TIME THIRTY TWO MINUTES FORTY FIVE SECONDS CAR PERFORMED PERFECT=

FRANK BRISKO.

9-24-32

TRIPLICATE—This copy is to be retained by entrant. It is your record of prizes and conditions. Keep for reference.

2762

OFFICIAL AAA ENTRY BLANK AAA SANCTION No.

An automobile race meet for cars and drivers in good standing with the Contest Board of the American Automobile Association will be held as described hereinafter.

~~LANGHORNE SPEEDWAY (Management Dick Dunn) SATURDAY, SEPT. 24, 1932.~~

Open to one and two-man cars, 366 cu.in. or under. Superchargers and riding mechanics barred. Cars must pass inspection by Technical Committee both as to condition and appearance before time trials. Cars not passing inspection will be automatically eliminated.

EVENT 1. FIRST ELIMINATION HEAT- 10 MILES. Field comprises first 12 cars in time trial positions. First 5 finishers qualify for Event 5.

EVENT 2. SECOND ELIMINATION HEAT- 10 MILES. Field comprises cars finishing 6th and later in Event 1; balance to be drawn from next fastest cars in qualifying trials for a total of 12 starters. First 5 finishers qualify for Event 5.

EVENT 3. THIRD ELIMINATION HEAT- 10 MILES. Field comprises cars finishing 6th and later in Event 2; balance to be drawn from next fastest cars in qualifying trials for a total of 12 starters. First 5 finishers qualify for Event 5.

EVENT 4. CONSOLATION RACE - 20 MILES. Field comprises cars finishing 6th and later in Event 3; balance of field to be drawn from next fastest cars in qualifying trials for a total of 16 starters. First 3 finishers qualify for Event 5.

EVENT 5. FIFTY MILE FINAL SWEEPSTAKES. Field comprises first 5 cars to finish Events 1, 2 and 3 and the first 3 cars finishing Event 4. Starting places determined by positions in time trials.

(DIVISION OF PRIZE MONEY)

EVENT 1. 1st $50; 2nd $35; 3rd $25. EVENT 2. 1st $45; 2nd $30; 3rd $20.
EVENT 3. 1st $40; 2nd $25; 3rd $15. EVENT 4. 1st $75; 2nd $40; 3rd $25.
EVENT 5. 1st $800; 2nd $400; 3rd $150; 4th $100; 5th $60; 6th $45; 7th $20. TOTAL PURSE $2,000.00.

Where duplicate times are made, precedence will be given car making earliest trial. Track available for practice Friday, Sept. 23, from 12 noon to 5 P.M. Credentials will be issued at track Sept. 24.

TIME TRIALS ON DAY OF RACE BETWEEN 10 A.M. and 2 P.M.

ENTRIES CLOSE AT MIDNIGHT, Sept. 20, 1932. RAIN DATE - Sat. Oct. 1,1932.

MAIL ORIGINAL BLANK TO: Dick Dunn, 252 So. Broad. St., Phone Pen.1010.

MAIL DUPLICATE BLANK TO: Mr. Jos. C. Dawson, 667 No. Broad. St.,Phila.

Following is entry for one car and driver in race meet above described: **Date:**........, **193**...

Driver........ **Address**........ (Street and city) **Reg. No.** ....

Mechanician........ **Address**........ (Street and city) **Reg. No.** ....

Name of Car........ **Model**........ (One or two-man) **Reg. No.** ....

Owner........ **Address**........ (Street and city)

Year Built........ **Weight**........ **lbs. Seat Width**........**inches. Painted: Color**........**Number**........

Bore........**inches. Stroke**........**inches. No. of cylinders**........ **Piston displacement**........**Cu. in.**

It is hereby understood that the car described above and driver named are to appear at the above-mentioned meeting, except as provided in the rules of the Contest Board, and particularly Rules 99 and 217 thereof. The Official Competition Rules, and the approved supplementary regulations are hereby made a part of this contract.

Entrant's Signature........ **P. O. Address**........

Illustration 8-5
Langhorne entry form.

An absolutely splendid result, as reported by the Four Wheel Drive Auto Company's local agent. This was, after all, only the car's third outing and it was high in the money, such as the money was.

GU2 51 NL COLLECT=BRISTOL PENN 24

W A OLEN=
CLINTONVILLE WIS=

CAR PLACED SECOND IN TEN MILES EVENT AND SECOND IN FIFTY MILE EVENT IN FIRST CLASS CONDITION EXCEPT HEATING BADLY AT CENTER DIFFERENTIAL STOP BRISKO SUFFERING FROM BURNING LEG RAN WONDERFUL RACE COMING WITHIN ACE OF FIRST PLACE BOTH EVENTS OUR CAR ONLY PLACING CAR TO FINISH IN GOOD CONDITION=

JOHN GRAY.
911A

John Gray, a company man on the scene, wrote a stirring and quite triumphant report on the events at Langhorne that day. (See sidebar on pages 159 and 160.) As in practically all American racing cars of the epoch, fuel feed depended on air pressure within the fuel tank, maintained by a hand pump on the instrument panel. Brisko had to drive while fighting a leak in the system.

Joy reigned in Clintonville.

SEPTEMBER 26, 1932

FRANK BRISKO
PAOLI PENNSYLVANIA
CONGRATULATIONS AND BEST WISHES FOR CONTINUED SUCCESS
THE FOUR WHEEL DRIVE AUTO COMPANY
11:45 A.M.

A note from Jim Bell arrived, saying "Mr. Olen has newspaper clippings. This was a great race and our car was the class of the show."

Hanson's reaction:

September 27, 1932

Mr. Frank Brisko
3124 North 39th Street
Milwaukee, Wisconsin

Dear Mr. Brisko:-

Attached is a copy of a telegram which we sent you yesterday at Paoli, Pennsylvania.

We were certainly very glad to receive the news of your success and we hope that you will add more to this record.

If you can find the time to do so, we would appreciate your giving us particulars regarding this race. It may be, however, that you will be in Clintonville and at that time it might be more convenient for you to dictate your story to a stenographer. We would like this information for our records and also as material we can refer to next May in the event that we enter the Indianapolis Race.

Very truly yours,

THE FOUR WHEEL DRIVE AUTO COMPANY

W. M. Hanson
Advertising Manager

PS
Attachment.

Ted Miller to the Company regarding the returned U-joint parts: This cool rebuke seems to jibe with Four Wheel Drive Auto Company's letter of September 23rd to Miller (p. 139). The credit memo Miller refers to is not in the file.

September 27th, 1932

Four Wheel Drive Auto Co.,
Attn; Mr. B. V. Welch,
Clintonville, Wisc.

Gentlemen;

In response to your letter of the 23rd in which you state that you are returning a shipment consisting of 2- Universal joint blocks, and the inner & outer housing, please be advised that we will gladly pass you full credit when the shipment is received.

However, we believe that you are in error when you state that the shipment was delayed. Mr. Brisco telegraphed us that he was in need of these parts on the 14th, and same was delivered to the Express company on the afternoon of the 14th. This service, you should agree, would be difficult to better. One day service is hard to beat.

The credit memo will be promptly forwarded cancelling the charge for the goods which was made direct to Mr. Brisco.

Very truly yours,

HARRY A. MILLER, INC.

Ted Miller
TN

Brisko transmitted the Company's share of his winnings to Olen on October 3rd.

Clintonville, Wisconsin.
October 3, 1932.

Four Wheel Drive Auto Co.,
Clintonville, Wisconsin.

Gentlemen:

I drove your race car at a race in Detroit. There were 40 entries and only 14 starters. I was in 6th position at the end of the 67th lap when a universal joint gave way and it was necessary to take the car out of the race. In this race we drove into a very deep rut which threw the mechanic up from the seat and as he came down into the seat he broke three ribs. He was taken to the hospital and I there paid the hospital bill amounting to $35.00.

I then drove the car at Philadelphia on September 24th at the Langhorn Speedway. There were about 50 entries and 14 starters. I ended up in second place, making one lap at the rate of over 100 miles per hour on a dirt track. In this race I won $400.00 for second place and also $30.00 in another event, making a total of $430.00. I herewith remit $107.50.

Yours very truly,

Frank Brisko

Frank Brisko

Brisko's letter and check were received and all were acknowledged on the same day, along with reimbursement of hospital expenses.

October 3, 1932.

Mr. Frank Brisko:

I hereby acknowledge receipt of $107.50.

I am attaching $35.00 to reimburse you for the hospital expense of the mechanic at Detroit.

Yours very truly,
FOUR WHEEL DRIVE AUTO CO.

Walter A. Olen
President

W.A.Olen,
ELV

Service manager C.H. Redman to Brisko on oil consumption:

CC: H. G. ENGEL

October 5, 1932

Mr. Frank Brisko
3124 North 39th St.
Milwaukee, Wisconsin

Dear Mr. Brisko:

It is important that we have the following information, just as quickly as possible and would appreciate very much, your sending it back to me, by return mail.

Will you please let me know, the amount of oil that was put in the crank case of the racing car, before the race started and the amount of oil that was left in the crank case after running the full fifty miles.

Yours very truly,
FOUR WHEEL DRIVE AUTO CO.

C. H. Redman
Service Manager

CHR:I

Brisko replied in very legible handwriting two days later:

Milwaukee, Wis.
Oct. 7, 1932

Mr. G. H. Redman.

Dear Sir:

Received your inquiry of Oct. 5 & am very glad I can give you the desired information.

In regards to the oil in the crank case, I put a gallon of oil, as I had the pan off & the motor was perfectly dry, & when I drained it we had 1 1/2 quarts as you know this motor has a dry sump.

We had about 6 gallons of oil at the start of the race. But due to a leakage we lost about 2 gallons. The cover on the oil tank was very loose. Quite a few screws were lost. Undir normal conditions the motor should run 500 miles on 6 gallons of oil.

I am figuring on coming up to your factory Monday.

Yours Truly

Frank Brisko

F. Brisko

(new address) 1228 N. Hawley Road
Milwaukee, Wis.

Redman acknowledged this reply, with thanks for promptness.

October 10, 1932

Mr. Frank Brisko
1228 No. Hawley Road
Milwaukee, Wisconsin

Dear Mr. Brisko:

We have received your letter of October 7 in reference to the oil consumption on the FWD racing car.

We wish to thank you for the prompt answer relative to our inquiry, as this information is desired in the field.

Yours very truly,
Four WHEEL DRIVE AUTO CO.

C. H. Redman
Service Manager.

LAN

On October 11th, Brisko brought the car's transfer case to the Four Wheel Drive Auto Company factory for the purpose of discussing its intolerable overheating with chief engineer H.B. Dodge. Dodge's letter of October 12th to Olen contains a valuable description of the problem and the band-aid treatment that was proposed. In this description, the badly made gear teeth do not sound like anything made by Miller and his shop boss, Offenhauser. They were herringbone teeth, which would have been a first for Miller, who I believe habitually purchased his very satisfactory gears from Boston Gear Company catalog stock. These very special herringbone gears may have been from an offbeat source. From this description we can also see that, while the transfer case remained a dire problem, Brisko seems to have eliminated a simple but serious clutch defect.

October 12, 1932

Mr. Olen
General Manager:

Mr. Brisko was here yesterday and brought up the transmission and transfer case assembly off our racing car. We have gone over all parts and find the bearings become very hot, and are discolored on account of heat. Mr. Brisko can obtain new bearings without cost, and will, therefore, arrange to send us a complete new set of ball bearings. The differential showed some signs of wear and had become somewhat discolored; however, no parts were scored and bearing surface is in first class condition. It will, therefore, not be necessary to replace these parts. Also the gears are entirely serviceable. The gear teeth, however, were very poorly machined and showed very poor workmanship.

Mr. Brisko brought the clutch plate along and had the tool room spot weld the retaining nut in place. This nut backed off during the last race and made the clutch unoperative. This is one complaint of the Miller clutch; however, be locking the nut by spot welding, this trouble should be entirely eliminated.

We propose to make no change in the gears or gear case; however, we will install a 5-gallon tank of oil to lubricate transmission and transfer case, and will arrange to pump cold oil in between the gears so that the gears are positively lubricated at all times with cold oil, and will install another pump to draw out the hot oil. I believe this method will relieve the trouble they had due to excessive heat, and the car will then be in perfect racing condition.

To make this change it will be necessary to obtain a dual pump in connection with an auxiliary oil tank, which should have fins or some means for cooling oil. Also necessary piping from oil tank to pump, and gear case. The cost of the new parts will possibly amount to $200 including the labor of installing them.

I am attaching hereto a rough sketch showing briefly the proposed oiling system for gear box.

Kindly advise if you wish to have this work done as above outlined.

Yours very truly,

H. B. Dodge,
Chief Engineer.
LAN
Attach.

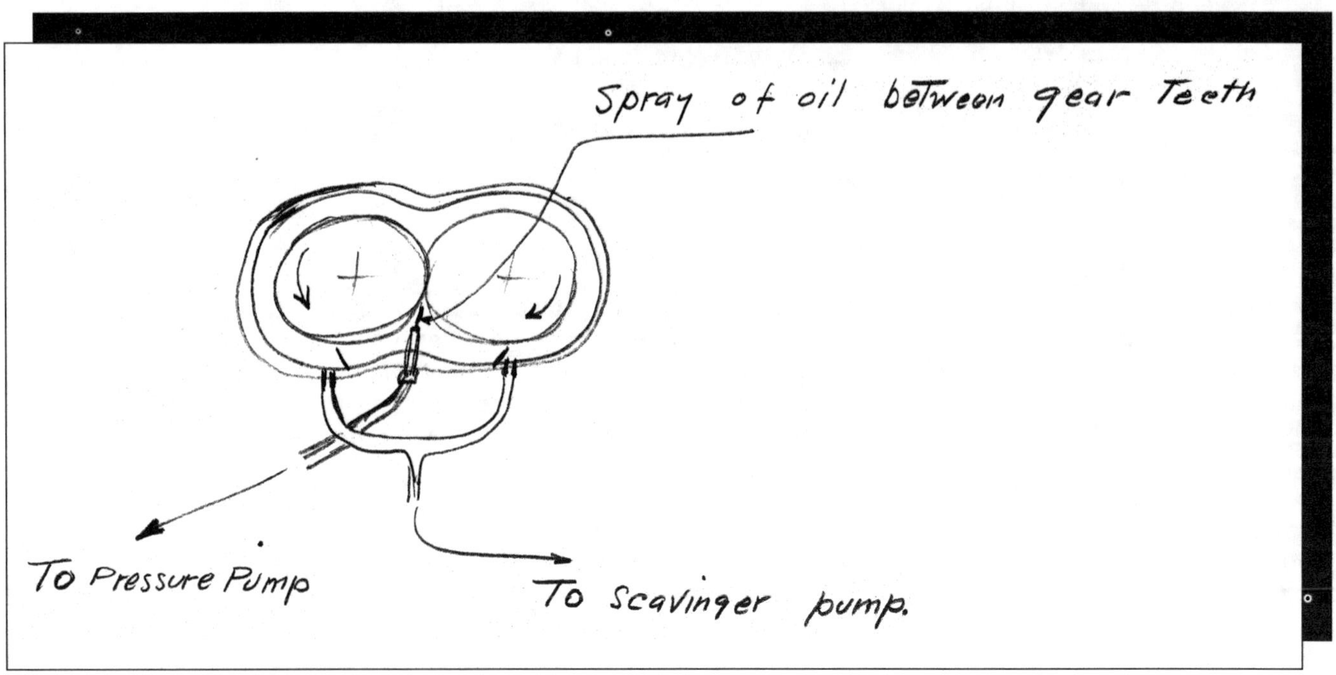

Olen, signing himself general manager instead of president, referred Dodge to Harry Miller, whom he associated with engineering "of the highest type." And he alluded to a grandiose new exploitation of the car. Brisko had revealed its potential, and the Company's technical and sales people were agog.

October 13,1932.

Mr. Dodge,
Chief Engineer.

Replying to your letter of the 12th with reference to the special oil pump for the racing car, I think this method of remedying the trouble should be submitted to Harry Miller before it is used. The engineering of this car should be of the highest type because it is a sample car that eventually will be shown all over America to its finest engineers and mechanics.

Yours truly,

Walter A. Olen
General Manager.
WAO:EL

Dodge to Miller, following Olen's instructions.

October 14, 1932

Mr. Harry A. Miller
6233 Gramercy Place
Los Angles,California

Dear Mr. Miller:

Mr. Olen has requested that designs be prepared covering a new transfer case for our racing car. This gear case to be larger in diameter to give more clearance between the diameter of the gear and the inside of case; also the outside of gear case to be ribbed so that it will radiate heat readily.

Will you, therefore, kindly send us detailed drawings of the present transfer case and in the event you have prepared new drawings covering the ribbed construction, we would be gland to receive them. If we have your original drawings, it will save a great deal of time in measuring up the present parts inorder to obtain accurate machined dimensions.

We will be glad to hear from you advising us fully just what should be done to this case inorder to eliminate the heating trouble experienced in the past, and will be very glad to receive your recommendations.

Yours very truly,
FOUR WHEEL DRIVE AUTO CO.

H.B. Dodge,
Chief Engineer.

LAN

The rapport between car owner and campaigner was close. Brisko, presumably to Dodge, in his tidy longhand:

Milwaukee, Wis.
Oct. 14 - 32

Dear Sir:

I have the bearings & oil pump at your branch here in Milwaukee. But I would advise you to wait with the oil pump as I will be down to Clintonville Monday & will bring oil pump gears along also a case which is better suited for the work, we want. We can drive this pump direct & that will be less moving parts. In case you won't be in town Monday kindly let me know.

Yours truly

Frank Brisko
1228 N. Hawley Rd.

Dodge, having followed Olen's directive, found a way five days later to submit to Harry Miller the proposal at which he and Brisko had arrived.

October 19, 1932

Harry A. Miller, Inc.
6233 Gramercy Place
Los Angeles, California

My dear Mr. Miller:

I wrote you on October 14 regarding what should be done to the transfer drive in our racing car inorder to eliminate excessive heat, and otherwise give the best possible results. Now that you have torn down the FWD which is a duplicate of ours, you may have discovered something which would indicate that certain changes in design are desireable.

I believe it has been your opinion right along that the transfer case should be made larger to give more room between gears and case, and that the case should be ribbed to radiate excessive heat. We would like to know if you believe this change is all that is necessary.

It would not be difficult to install a pump at the rear of transfer case. This pump to deliver a certain amount of cool oil in between the gears and also provide a scavenger pump to take away all hot oil which oil would pass through a radiator into a reserve tank. The reserve tank to be approximately 5-gallons capacity. By this means cold oil would at all times be forced in between gears and a certain amount of excessive heat would be carried away through the scavenger line. If the circulation of oil is great enough, ther e would be very little heat to be radiated by the case itself.

We would also like to have your opinion regarding whether or not it would be adviseable to install the oiling system above outlined without changing the gear case. Quite possibly the cooling action through circulation of oil would eliminate the necessity of any case patterns.

During the race at Detroit and Philadelphia driven by Mr. Brisko, the case became very hot and burned Mr. Brisko's leg quite badly. He was also compelled to slow down due to the excessive heat. This heat developed after the race had progressed a very few miles.

In the event you recommend a change in the oiling system as above outlined, would you then continue to use castor oil or to use a mineral oil? Quite possibly better results would be obtained by the use of the new type EP lubricants to lubricate these gears.

We would be glad to hear from you promptly regarding the alterations in the transfer drive, as we may decide to enter this car in another race this fall but would not enter car unless the heating trouble has been corrected.

Yours very truly,
FOUR WHEEL DRIVE AUTO CO.

H.B. Dodge,
Chief Engineer.

LAN

Ted Miller's response to Dodge's first letter on this subject. (The blueprints referred to in this letter were not in the file.)

October 19th, 1932

Four Wheel Drive Auto Co.,
Clintonville, Wisc.

Attn; H. B. Dodge, Chief Eng.

Gentlemen;

This will acknowledge receipt of your letter of the 14th which requests detailed drawings of the transfer case on our four wheel drive race jobs.

We are enclosing herewith detailed blue prints of the transfer gear housings, both front and rear halfs. We are also sending an assembly print showing transfer gears. We would like to call your attention to the fact that we have cut two additional grooves in the gears to allow the oil to escape.

The reason that the case became hot was on account of the friction of the oil. However, we believe that this will eliminate the heating difficulties.

We are enclosing also a sketch of a rear engine mounting. You will note that this gives a three-point suspension. It was our plan at first to make the jobs like this when we constructed them, but changed and now have decided to make the change definitely.

Your comments will be gladly received and if we can be of any further assistance, do not hesitate to write us again.

Very truly yours,
HARRY A. MILLER, INC.

Ted Miller

Ted Miller
TM
enc.

Ted Miller's reply to Dodge's second letter. The oil did leak out massively, according to Goossen.

October 25th, 1932

Four Wheel Drive Auto Co.,
Clintonville, Wisc.

Attn; H. B. Dodge

Gentlemen;

Your letter of the 19th has been received and we are of the opinion that the question of the transfer gears overheating was covered by our last letter, dated the 19th, in which we explained that we will cut the two oil clearance grooves, allowing the oil to escape more freely, thereby, obviating the heating of the oil.

We do not hardly believe that the oil pump is necessary as this would complicate matters requiring a cooler, etc. There will, of course, always be plenty of oil in the gear housing as it cannot leak out.

We are also enclosing a blue print of the flexible coupling which is incorporated just ahead of the rear axle housing. This was found necessary when we mount our motor at three points as there will be a certain amount of movement to the unit.

We regret the delay in answering your letter but same was mislaid and only discovered this morning.

Very truly yours,
HARRY A. MILLER, INC.

Ted Miller

Ted Miller
TM
enc.

The cross fire continued as Dodge, unconvinced, replied to Ted's first letter.

October 25, 1932

Harry A. Miller, Inc.
6233 Gramercy Place
Los Angeles, California

Dear Mr. Miller:

We have your letter of October 19 inclosing drawings covering transfer case parts for our racing car, and note your recommendations which we understand are remachining gears, adding grooves to permit oil to escape. Also a new motor support which appears much larger in diameter, which also requires a new frame cross member.

We understand these two changes constitute the only improvements you believe necessary; however, your letter undoubtedly was written before you received our letter of the 19th, in which letter we asked your opinion regarding a special oiling system for the gear case. We would like to hear from you regarding your opinion of this system.

The proposed method of grooving teeth would, undoubtedly, permit oil to escape and reduce oil pressure between gear teeth. However, the gears in our car indicate that the pressure has been quite excessive and the teeth show signs of hammering out due to extreme tooth pressure, and we are wondering whether or not the gear teeth will be weakened by grooving sufficiently to cause trouble from another source.

We have not had any trouble that I know of due to the motor suspension and would like to know if you believe a new suspension would be advisable, and should be installed on our car, or whether this is an improvement which would only be installed on any cars built in the future.

We will be glad to hear from you as quickly as possible, as we may decide to enter this car in a race this fall, and would not want to do so unless it is in the best possible condition.

Yours very truly,
FOUR WHEEL DRIVE AUTO CO.

H.B. Dodge

H.B. Dodge,
Chief Engineer.

LAN

Ted's concluding statement in this exchange of correspondence still ignored the Dodge-Brisko proposal.

October 31st, 1932

"VIA AIR MAIL"

Four Wheel Drive Auto Co.,
Clintonville, Wisc.

Attn; Mr. H. B. Dodge, Chief Engineer

Gentlemen;

This will acknowledge your letter of the 25th and with reference to the new engine support which we are installing on our Four-wheel-drive car, we believe this design to be superior allowing for a three-point support of the motor, although, it is not absolutely necessary for you to do this on your job. Our thought was to keep you informed of improvements as we make them from time to time.

If you prefer not to groove the gear teeth because of weakening them, it would be satisfactory to drill holes in the tooth spaces to allow the oil to escape. One of the transfer gears is casehardened, therefore, it will not easily be drilled, while the other gear which carries the differential is heat-treated and could be drilled without any trouble.

Any further questions, or comments or criticisms, will be welcomed.

Very truly yours,
HARRY A. MILLER, INC.

Ted Miller

Ted Miller
TM

Dodge took his time—about two weeks—to pass the entire transfer case discussion to Olen. The General Manager-President ordered, "Have the car fixed and brought back here."

November 16, 1932

Mr. Olen
General Manager:

I am attaching hereto correspondence from Harry A. Miller, Inc. which letters were written by Ted Miller. This in regard to changes to be made in our racing car.

From these letters you will note that the auxiliary oil cooler for gear box is not believed necessary and the only change they suggest is that the gears be grooved to permit oil to escape between gear teeth readily. However, this would weaken the gear materially and their latest suggestion is that holes be drilled at the bottom of teeth instead of grooves.

This car is now torn down and I believe we should drill gears as recommended and reassemble same so that it will be in operating condition.

Yours very truly,

H. B. Dodge,
Chief Engineer.
LAN
Attach.

[handwritten note at bottom of letter] Have the car fixed and brought back here. WAO

On the same date that Dodge referred the dossier to Olen, Harry Miller had entered his own four-wheel-drive car in a 150 miler on the fine banked dirt track that existed then in Oakland, across the bay from San Francisco. Happily for posterity, the Four Wheel Drive Auto Company's local representative observed and made this very competent report to the home office.

COPY

[To] S. F. Wick [at] Factory 11/16/32

[From] K. O. Stockland, Jr. [at] San Francisco Branch

Dear Stanley:

I went over to the Oakland track Sunday to see the 150 mile dirt track championship race and more especially to watch the Miller FWD which Miller had entered. Louie Meyers was driving the car. He was national champion in 1928 and 1929.

The track is a one mile oval and conceded the best dirt track in the country. Fifteen cars were allowed in the race, the FWD being one of those that qualified. Our car caused quite a sensation especially before the race during the warm-up runs when the track was not crowded and Meyer could let it out.

The race started off with a bang, everyone driving like wild men to try and gain a few laps for the long grind. Meyer hit a steady pace holding to a speed of as near a lap in 38 seconds as possible and paying no attention to the others. Miller's theory was that at the pace they were driving that other cars would be in for tires and would develop mechanical trouble and when this happened and the traffic thinned out, that Meyer could then let her out and as he felt we could easily go thru without a tire change that we had a good chance of coming in the money. Unfortunately on the 24th lap, right in front of the pits, as near as we can find out, the cap screws sheared off the flywheel, locked all wheels and threw the car into a spin, wiped off the inside track rail with the right two wheels (the cars were running counterclockwise on the track) mashed up the tail of the car and came to rest on all four wheels, facing again in the proper direction. As this put us definitely out of the race, the car was towed into the pit.

In talking to Miller he said he was afraid that the flywheel might come loose and had intended putting dowels in it to take care of the shear but the car was assembled before this was done and they took a chance.

Meyer was very well pleased with the way the car handled and said he felt entirely safe at all times. However, they both agreed that the car was geared too high for this track and thought also that it was a little too heavy on a mile stretch. There being no straight away to amount to anything, it is necessary to slow down and pick up constantly, especially with many cars on the track and with its weight and the Indianapolis ratio in it, it didn't have the pick up it might have had with a little lower ratio.

I had hoped to be able to wire you Sunday that we had won a race but such is the fortunes of racing. Meyer drove a good intelligent race while he was in there and I look for big things from this car if and when Miller enters it again.

Yours truly,
(signed) M.O. Stockland, Jr.

We do not know the date of Olen's order to return the Company's car to the factory. Dodge passed the message to Brisko ten days after having submitted the matter to Olen.

November 25, 1932

Mr. Frank Brisko
1228 No. Hawley Road
Milwaukee, Wisconsin

Dear Mr. Brisko:

We are sending to our Milwaukee Branch all the transmission parts you left here; also the box of parts covering a special pump which parts are your property.

We have drilled one of the gears to permit oil to escape. This in accordance with directions given by Mr. Miller.

Will you kindly call for these parts and reassemble the car; also would like to have you deliver it to Clintonville as promptly as possible. We would like to display this car and, therefore, would like to have it put in running condition without further delay.

Yours very truly,
FOUR WHEEL DRIVE AUTO CO.

H. B. Dodge,
Chief Engineer.

LAN

Brisko replied in his usual prompt and impeccable longhand style:

Milwaukee, Wis.
Nov. 29-32

Mr. H. B. Dodge
Dear Sir:

I received your letter of Nov. 25, & I will re-assemble the car as soon as I receive the parts, & I will deliver the car to Clintonville. I was asked to go to the Kissel Co. With the car quite a while back, but as the car was not yet ready, could not do so. Should you want me to go there now, please let me know by wire.

Hope to have everything ready & will be there as soon as possible.

Yours truly

Frank Brisko
Frank Brisko
1228 No. Hawley Rd.

We do not know the nature of the interest of the Kissel Motor Car Co., of Hartford, Wisconsin, in the Four Wheel Drive Auto Company's Miller. It could have been related to use of the four-wheel drive principle in a road car. Kissel was involved with the Ruxton passenger car, a pioneer of front-wheel drive.

December 1, 1932

Mr. Frank Brisko
1228 No. Hawley Road
Milwaukee, Wisconsin

Dear Mr. Brisko:

We have your letter of November 29 and wish to advise you that the racing car parts are now at the Milwaukee Branch, unless you have called for them, and note that you will reassemble car and return it to us promptly.

In this connection, it will not be necessary for you to call at Hartford and show same to the Kissel Company.

Yours very truly,
FOUR WHEEL DRIVE AUTO CO.

H.B. Dodge
H.B. Dodge,
Chief Engineer.

LAN

A very gingerly and delicately composed letter from Dodge to Harry Miller, stating that his remedy for the transfer case was ineffectual. This comes out chillingly in the third paragraph. There is no evidence of a reply to this letter.

December 9, 1932

Harry A. Miller, Inc.
6233 GramMercy Place
Los Angeles, California

Dear Mr. Miller:

We understand you have repaired your Four-wheel drive racing car and have entered it in races. We would be very much interested in knowing what improvements you made on this car, and wherein it differs from our car. We would also like to know what trouble developed during races; particularly in respect to heating of gear box. We would like to know if the gear box ran sufficiently cool during race to cause no alarm or whether excessive heat was developed.

You suggested certain changes in gears to prevent oil lock between the gear teeth. In the soft gear we drilled three holes at THE bottom of each tooth to permit oil to escape; however, we were unable to drill the hardened gear and, therefore, turned one groove in the center full depth of tooth. We were somewhat skeptical regarding turning two additional grooves as you suggested, as we feared this would weaken the gear to quite an extent; however, if you have machined your gears in this manner, we would like to be so advised, and would like to know what results you obtained.

With our present arrangement we towed this car a distance of eleven miles at a speed of approximately 70 miles per hour, and the gear box got so hot that it was dangerous to sit in the seat beside it; - this with no power being transmitted through the gears. It, therefore, seems as though some other change must be made in order to remedy this trouble.

We trust you will advise us fully regarding improvements on your car, and what benefit you have received from such improvements.

Yours very truly,
FOUR WHEEL DRIVE AUTO CO.

H. B. Dodge,
Chief Engineer.

LAN

On the same date that Dodge wrote to Harry Miller for new ideas for curing the transfer case problem, he wrote to noted boat-builder and racer Gar Wood. We do not have a copy of that letter. Wood's reply was enlightening, but off-trail. Perhaps if he could have observed the transfer case he would have recognized the cause of the trouble on sight.

December 21, 1932

Mr. H. B. Dodge
Four Wheel Drive Co.
Clintonville, Wis.

Dear Sir:

Your letter of the 9th has not been answered sooner due to my absence from the city. I note what you say regarding excessive heat generating in the gear transmission of your four wheel drive racing car. I recall that Mr. Miller asked me for a suggestion as to how to overcome this heating condition, and I believe I made some suggestion at that time. However, since then we have learned a great deal about running gears supported on ball bearings at excessive speeds, and we find that most of the heat is generated due to the ball bearing action on the oil. Apparently the compression of the oil due to the high speed of the ball, causes a very rapid increase in temperature. To overcome this, we ran our gear boxes transmitting 3400 h.p., practically dry. The gears were three pitch, three and a half inch face, one five inch diameter and the other 15 inch diameter, the small one turning approximately 8000 r.p.m. For lubrication we used nothing more than a fine vapor so that the metal parts were just moist, which apparently was sufficient lubrication. I would suggest that you try this method. I don't believe you need very much oil for lubrication. It seems to me our discovery of this condition may in the future lead to a different means of lubricating high speed machinery.

I am not familiar with the detail of your design, but I believe you can make such alterations as is necessary for making a test of this kind. A small amount of oil is better than a great deal, properly placed. You may have to develop some method for spraying the oil into the gear box, then draw it off the bottom. A small gear pump could be used for this purpose and suitable spraying nozzles could easily be designed. I believe the large size nozzles used in oil burners on the market now would do very well. If you should try this method of lubrication, I would be pleased to hear of the results of your tests.

Very truly yours,

Gar. Wood

Gar Wood

In the second paragraph of Dodge's next communication to Olen on this subject, the engineer put his finger on the exact cause of the problem. Dodge suggested waiting to hear from Harry Miller, but Olen dashed off the words "go ahead." On this progressive note the Four Wheel Drive Auto Company concluded its first year of apprenticeship in the racing business.

December 22, 1932

Mr. Olen:

I have written to Gar Wood and to Mr. Miller regarding changes to be made in our present gear box in racing car, and have not as yet heard from Mr. Miller, but have a letter from Mr. Wood dated December 21. You will note he recommends that provision be made to feed a small amount of oil on gears by a spray, and that any excess of oil be drawn out of case so that it will not accumulate. This is in line with tests we have been conducting and have obtained the best results from this method.

The inside of the gear case is very uneven, and on certain locations the wall comes in so that it nearly meets gears allowing no space for oil or spray. I believe we should make new patterns and make the case larger inside and as smooth as possible. All ribs would be located on the outside instead of inside as at present. However, before doing anything I believe we should await a reply to our letter to Mr. Miller, as he may have some valuable experience with his racing car.

Yours very truly,

H. B. Dodge,
Chief Engineer

LAN

[handwritten note at bottom of letter]
Go ahead
WAO

Brisko shows his value

September 25, 1932

Mr. W. A. Olen
Clintonville, Wis.,

Dear Mr. Olen:

In the night letter sent you last night I tried with out [sic] running your bill up too high to give the events of most import in connection with the race run at Langhorn [sic] with the FWD car.

I arrived at the track too late to watch the qualifying run but was very pleased to see that Mr. Brisko had drawn third position at 38 seconds flat as against Wild Bill Cummings time of 37.2 and Deacon Litz in 37.3 for the mile on dirt track which I am sure you will agree is pretty fast time. The first six qualified in less than forty seconds.

In the first ten mile event Brisko dropped out after a few fast lapspleading [sic] a slipped carburetor spring, this alibi placed him at the pole position for the second ten mile event and eliminated four fast cars against him. This was a fine piece of strategy in that the first event turned out to be a heart breaker and terrible hard on the three leading cars numbers ten sixteen and fourteen. These cars finished 16 first 10 second by a half car lenght [sic] and fourteen within a length of 2nd. Time 10 miles 6min-24 and 2/5 seconds.

The second event with Brisko at the pole was a bit slower but a very pretty race. Brisko led the field until the ninth lap when be lost his air pressure and his lead for nearly the lap. He finished handily in second by some fine driving with one hand while pumping frantically to keep going. Time 6 min. 39 and 1/5 seconds for ten mile.

The third event was much slower and except for qualifying some cars for the sweepstake was not interesting as we were not in.

During the running of the third and fourth events Mr. Brisko, and his man looked high and low for the remedy to the air leak and some way to keep from burning his leg on the center differential, which was heating badly. We placed a board about a 1/4 or 3/8 inch wrapped in cloth between his leg and the case, and smeared hard grease over the gas tank gaskets to seal th [sic] air leak.

The sweepstake of fifty miles started with Brisko in third position. On the first lap he let the 14 car pass him dropping to 4th and in the next lap let Arnold in 27 pass him. Driving a very careful and fine race he maintained the 5th position and let Cummings and Litz in 10 and 16 cars run their heads off in a dual [sic]. The first six laps were run in 3.7, the first 20-mile in 11.32. On the 20th lap the fast pace told on Litz in 16 and he dropped out moving following cars up with the FWD in fourth. At thirty miles Brisko moved up to challenge Arnold in 27 and passed him. At the thirty-five mile after Arnold had passed Brisko only to be passed again Arnold dropped out with a stone bruise over the eye. This moved us into third place. At the 40 mile which was run in 26.10, Brisko moved up to worry and challenge Patterson in 14. Still running a smart race in that he worried Patterson into frequent changes of pace Brisko played up to the 47th when he moved into second place. Cummings in 10 had run a spectacular race haveing [sic] lapped the field and was leading Brisko by a lap. He had however given his car an awful beating and now his motor was missing frequently and had slowed up noticeably. The third place car No. 14 ran into difficulty with motor trouble when Cummings finished ON A Flat Time in 32 minutes45 and 2/5 seconds. Brisko finished very shortly behind him coming in FAST and then Patterson coasted in for a third with his motor dead. Arnolds 27 finished limping.

Of the cars placing the FWD was the ONLY ONE TO FINISH IN GOOD SHAPE AND COMING IN FAST. Had the race been another mile and a half longer there is no doubt of our finishing first. This is the outstanding part of this race. And only 9 of 18 starters finished race 5 of the 9 limping in with difficulty.

The FWD finished fine but Brisko certainly exhibited vast store of real courage in staying with it to place as he did. The cockpit was so hot that I couldnt [sic] put my hand down in even after it was load ed [sic] on its trailer. Brisko had burned his leg at Detroit badly the same way and had a thick bandage on. Getting out of the car his trouser leg was burned and his sock and bandage badly scortched [sic]. The board was scorched as well as the cloths we had wrapped around it.

Considering the fact that the car handles so differently in and out of the turns and the condition Brisko was faceing [sic] in the hot seat he ran a wonderful race.

The FWD excited the entire crowd and had the crowd pulling for it. The absence of slippage on the turns and its great get away on the straightaway were demonstrated very clearly. On the back stretch coming out of the turn the FWD fairly flew away from any and all of the cars in the race. We lost considerable speed going into the turns due to the difference in handling. Very favorable anouncement [sic] was made to the crowd through the loud speakers on the fact that the car 58 was the only 4 wheel drive in the world, etc.

The clipping from the Philadelphia Inquirer and my program is attached. Ed Bittner of Bell Telphone [sic] took some motion picture of the race about the time we challenged for second place. If they are good I will get them for you.

Yours very truly

A still excited and trembly - -

J. W. Gray

Johnnie [John Gray]

Chapter 9

Perhaps the Fastest Car in the Indy Field

Illustration 9-1 Frank Brisko in 1933 with riding mechanic Johnny Hughes. (RMRC)

This series of adventures began on April 24th, 1933, with the signing of a new contract between the Four Wheel Drive Auto Company and Frank Brisko. The company leased the car to him specifically to drive in the Indianapolis 500. This lease had much in common with the previous contract between the same parties, with a particularly notable difference: The Miller name was deleted from the name of the car. There was no correspondence from Brisko; he had more-important things to do.

THIS INDENTURE made and entered into this 24th day of April, 1933, between the Four Wheel Drive Auto Company of Clintonville, Wis., hereinafter called "Lessor" and Frank F. Brisko of 1228 North Hawley Road, Milwaukee, Wis., hereinafter called "Lessee".

WITNESSETH:

That the Lessor hereby leases to Lessee and Lessee hereby agrees to rent from the Lessor, one racing car described and known as the "Four Wheel Drive Special" now at the factory of the Four Wheel Drive Auto Company for the term beginning April 24th, 1933 and ending on June 3rd, 1933 at 5 o'clock P.M.

It is expressly agreed and understood that said racing car may be driven by lessee in the 500 mile automobile race to be run at Indianapolis Speedway on May 30, 1933, and that lessor will furnish all repair parts, lubricating oil, and tires required during the term of this lease, to put and keep said car in condition and to participate in said lease.

IN CONSIDERATION WHEREOF, the lessee binds himself to pay as rent for same a sum equal to 50% of the gross prize winnings or earnings of said car earned as a result of the lessee participating in said race, said sum to be due and payable to lessor at the office of the Indianapolis Speedway immediately upon payment of said prize money or earnings.

The lessee further agrees that immediately after the 500 mile automobile race on May 30, 1933 at the Indianapolis Speedway, he will deliver said Four Wheel Drive Special car to the Four Wheel Drive Auto Company at Clintonville, Wisconsin, said delivery to be made at the point designated not later than 5 o'clock P.M. June 3, 1933, unless otherwise agreed upon in writing, between lessee and lessor.

It is further agreed that lessor shall not be liable to said lessee for damages of any kind or from any cause whatever to the persons or property of lessee, or any person claiming by, through or under him, and lessee agrees to defend at his own expense all damage suits which may be brought upon claims relating to said cause or the use thereof by the lessee.

IN WITNESS THEREOF, the parties hereto have hereunto set their hands and seals the day and year first above written.

FOUR WHEEL DRIVE AUTO CO.
Lessor

BY Walter A. Olen
President

Attest:

Frank Gause

Frank Gause

Secretary

Frank Brisko

Frank Brisko
Lessee

Attest:

Frank Gause

Frank Gause
Secretary

The company file contains a blank entry form for the 1933 "Sweepstakes," the closing date for which was May 1 at midnight. It is interesting to note that, since the previous year, the form had been updated to include four-wheel drive.

TWENTY-FIRST ANNUAL INTERNATIONAL SWEEPSTAKES

An automobile race for cars and drivers in good standing with the Contest Board of the A. A. A. will be held as described hereafter, under OFFICIAL SANCTION No. 2850

DISTANCE 500 MILES

TUESDAY, MAY 30th, 1933, 10:00 A. M.

CLASS E

NON-STOCK, 366 (6,000 CC. OR 6 LITRES) CUBIC INCHES PISTON DISPLACEMENT OR LESS—SEE APPROVED SUPPLEMENTARY CONDITIONS FOR WEIGHT LIMITS

First Prize	$20,000	Sixth Prize	$ 2,200
Second Prize	10,000	Seventh Prize	1,800
Third Prize	5,000	Eighth Prize	1,600
Fourth Prize	3,500	Ninth Prize	1,500
Fifth Prize	3,000	Tenth Prize	1,400
		Total	$50,000

In addition to above prize monies a consolation prize will be awarded, the amount and manner of distribution to be announced on or before May 30th, 1933.

The management reserves the right to increase total prize monies to be distributed by the addition of not less than 75% of the entry fees collected, as provided by Rule No. 2 of the Approved Supplementary Regulations.

TROPHIES
PREST-O-LITE TROPHY

Under the Management of

INDIANAPOLIS MOTOR SPEEDV
CORPORATION

E. V. RICKENBACKER, President T. E. MYERS, Vice-President E. D. MOORE, Treasurer

OFFICIAL ENTRY BLANK

INDIANAPOLIS MOTOR SPEEDWAY CORP.
INDIANAPOLIS, INDIANA

Enclosed is remittance of $200.00 entry fee for one car in race meet described herein.

Date________________19____

Name of Car________________Model and Year________________

No. of Cylinders__________Bore in Inches__________Stroke__________Inches

Piston Displacement__________Cu. In. Supercharger?__________Weight__________

Front, Rear or 4-wheel Drive?__________Four or Two-Cycle?__________

Gasoline—Diesel—Semi-Diesel?__________A. A. A. Car Registration No.__________

Name of Owner____________________

Address of Owner____________________

Name of Driver________________A. A. A. Registration No.__________

Address of Driver____________________

Name of Mechanician________________A. A. A. Registration No.__________

Address of Mechanician____________________

It is hereby understood that the car described above and driver named are to appear at the above-mentioned meeting, except as provided in the rules of the Contest Board, and particularly Rules 99 and 217 thereof.

All rules of the Contest Board, A. A. A., and the approved supplementary regulations are hereby made a part of this contract.

Signature of Entrant____________________

P. O. Address____________________

Entries Close Midnight, May 1, 1933

THIS COPY FOR CONTEST BOARD, A. A. A.—ENTRANT, DO NOT DETACH

Illustration 9-2 Entry blank, front, and rear.

As race time approached the Company released a fact sheet, combined with assorted product testimonials. We are already familiar with the latter, which were culled just after the 1932 race. The photo and caption obviously pre-date the new nix-on-Miller policy. We learn from the caption that F.E. Deupree, presumably a relative of powerful publicist Herman Deupree, was the Company's distributor for Indiana.

At the Indianapolis Race Track: The Miller FWD Special that qualified at 113.279. M. P. H Standing back of car left to right: H. B. Dodge, Chief Engineer FWD Auto Company; Barney Oldfield; Walter A. Olen, President and General Manager FWD Auto Company and Frank Gause, Secretary FWD Auto Company. Seated: BobMcDonogh, driver, and F. E. Deupree, Indiana distributor of FWD trucks.

The MILLER FWD SPECIAL has a wheel base of 100" and tread of 56". Fully equipped it weighs 2,325 lbs. The power plant is a V eight-cylinder with a 3½" bore and 4" stroke, giving a piston displacement of 308 cubic inches. It develops three hundred horse-power at 5,500 revolutions per minute. The including angle of the V is 45 degrees, making an exceptionally narrow engine.

A five-bearing counter balanced crank shaft is used. Side-by-side tubular connecting rods, four overhead cam shafts, two magnetos, and four down draft carburetors attached directly to the top of the cylinder block are used. The weight of the engine is 540 pounds.

The drive to the front and rear wheels is accomplished by means of propeller shafts from the transmission to the front and rear wheels. In the transmission there is a center differential which compensates for the difference in distance traveled between the front and rear wheels. It is obviously as essential at this location as the differentials in the axles. This differential is so arranged that it can be locked as desired, resulting in a positive drive from the transmission to the axles.

On the front axle the outboard joint is of the trunion and ring type, so placed that its axis is on the steering knuckle center. The steering knuckle pin is inclined so that a line extended through its center will coincide with the center line of the tire at the ground.

The rear axle shafts have bronze block universals, both inboard and outboard. Two chassis springs are used per wheel, and are quarter-elliptic, mounted one above the other. There is no rigid member between the wheel or differential case, instead a dead axle connects the two wheels, both front and rear, thus the wheels become semi-independently sprung.

Among the advantages resulting from four wheel drive in a racing car, passenger car or truck, are the following: No wheel spinning as maximum torque of engine is applied to all four wheels, better traction on oily or slippery tracks or roads, increased tire life as maximum stresses on tires are reduced, superior acceleration or get-away, turns or curves on tracks and roads can be taken at higher speeds with the greatest safety due to entire elimination of rear-end whip.

What Drivers Say About the FWD Racing Car!

I would like to express myself regarding the action of the Miller FWD Special on the track at Indianapolis. I found absolutely no slippage in going into a turn and pulling through as compared with a rear drive or a front drive at comparable speeds.

As you know, there are some rough spots on the Indianapolis track and I found that there was absolutely no slipping as experienced in other types of drives.

I had the car out one day during a rain and found that it handled the same on a wet track as well as a dry. In both cases we got up to a straight-away speed of 125 miles an hour. I appreciated the opportunity of driving the car and enjoyed the experience very much and if you decide to race this car next year, I would like to put my bid in for it.

Robert H McDonogh

For fifteen years, I have been driving race cars of all types, but since having driven the four wheel drive at Indianapolis this year, I sincerely believe their handling is incomparable. There is no sluing or skidding going in or coming out of the turns.

Too, the acceleration is so much smoother since the power is being transmitted to four wheels simultaneously, which is not done on any other type drive. The same is true of deceleration.

I firmly believe in the principle of the Four Wheel Drive.

Gus Schrader

I have never ridden in any car with the feeling of safety which I had after the first lap in the Four Wheel Drive job. The ease of taking the turns in this car at a higher rate of speed and with absolutely no skidding has sold me on the Four Wheel Drive principle in a far greater way than the four wheel brakes impressed me several years ago.

I must say that the car is beautifully designed and finished throughout and that I was especially impressed with the simplicity with which the drive to all four wheels was accomplished.

Ernest Olson

I was very proud to have managed your four wheel drive racing car in the last Indianapolis race. It is my idea of a racing car. I drove your car in a practice spin and it handled wonderfully.

Wishing you all the success in the world.

You know me
Barney Oldfield

Illustration 9-3 Fact sheet.

In May of 1933, the Company's monthly newsletter naturally was keyed to the race. Its cover was dominated by a broadside photo of the car, in its new livery, from which all Miller identification had been eliminated. This policy was adhered to throughout the publication's text. As we have seen, there was abundant cause for this, most if not all of which was brought on by Harry Miller's desperate financial situation. Perhaps the worst thorn in the Company's side was the existence of a car identical to its own, which would be compared and confused with its own. The duplicate car could detract from its sister car if it performed better or if it did the opposite. It was a permanent threat, over which the Company had no control whatsoever. Harry Miller had probably reasoned all along that the Four Wheel Drive Auto Company could not afford not to buy the duplicate car.

The confidence that Olen had shown in Brisko was vindicated mightily during qualifying trials at the Speedway. The fastest combination in the entire field was Wild Bill Cummings with a rear-wheel drive 8C270 Miller. Cummings averaged 118.531 mph for the required 25 miles. Brisko, in his four-wheel drive, was the second fastest of all, at 118.388 mph. This was one of the finest performances of Frank's career. After it, he "pronounced the FWD race car in perfect condition" and the Company newsletter went to press.

THE FOUR WHEEL DRIVE AUTO CO.

MONTHLY NEWS LETTER

CLINTONVILLE, WISCONSIN, U. S. A.

Vol. 14, No. 5 PRINTED IN U. S. A. May 1933

TO RACE AT INDIANAPOLIS MEMORIAL DAY

This FWD Racing Car was produced by us in collaboration with the most experienced high speed engineer in the world, employing the FWD principle of Four Wheel Drive. This car was built for the purpose of studying problems of high speed operation in order that greater safety may be incorporated in our high speed FWD trucks, than in any other truck on the highway.

Illustration 9-4 FWD newsletter.

FWD TO RACE AT INDIANAPOLIS MEMORIAL DAY

The FWD four-wheel-drive racer will challenge forty-one other makes of cars in the annual five-hundred mile speedway classic Memorial Day, May 30, at Indianapolis. The FWD is the only four-wheel-drive racer to qualify for the five-hundred mile grind, winning the coveted first-line position at a qualifying speed of 118.388 miles per hour on the 2 1/2 mile brick oval (the three fastest cars in a 25-mile qualifying run are awarded the front row positions, a distinct honor for the FWD).

The FWD racing car has been completely revamped at the FWD factory during the past few months. The car was built last year and brought to the Indianapolis track only a week before the race, having never turned a wheel. It qualified at over 113 miles an hour, but was forced out of the race due to engine lubrication trouble. Under the guiding hand of Frank Brisko, well-known race car driver, the car was further tested at Detroit and Philadelphia on dirt tracks, winning second place in the Philadelphia race. The excellence gained on these races resulted in some changes in the car and it is now believed that all mechanical difficulties have been overcome. Mr. Brisko, who will also pilot the FWD entry at the Indianapolis race, has pronounced the FWD racer in perfect condition after the 25-mile qualifying run.

The FWD racer was built and is raced for the sole purpose of studying the problems of high-speed operation in order that greater safety at higher speed and more economy will be incorporated in our high-speed FWD trucks than in any other truck on the highway. The FWD Company has already gained priceless information that is already incorporated into its trucks from the experience gained on the Indianapolis Speedway, that puts the FWD in advance of all its competitors in the way of safety, speed and dependability. The car was designed by the most experienced, high-speed engineer in the world in collaboration with FWD engineers, using the efficient FWD principle of drive. At the race FWD technical experts will observe its action at high speed as compared with the rear-drive and front-drive racers. As if under a magnifying glass they will study its advantages on saving tires, its action in the curves and its advantages against sluing and skidding on a slippery or wet track. The observations of other race car drivers as to its four-wheel-drive advantage for high-speed work, its weight distribution advantages, spring suspension and oil temperatures, and other technical points will be observed.

The Indianapolis track is sometimes called the castor-oil track because it becomes so completely saturated with oil and grease that it becomes extremely slippery and dangerous at the turns. Under these most exceptional conditions FWD engineers will study every phase of the car's operation for the purpose of refining and developing FWD trucks and particularly the FWD six-wheel truck for high-speed, heavy-duty, long-distance hauling.

Comments the Indianapolis Times: "Experts admit Brisko has the fastest racing car ever seen on the Indianapolis tracks."

The Indianapolis News reports. "Its four-wheel-drive principle makes it easy on tires on the curves."

Pre-race photos of the bare four-wheel drive chassis revealed a transfer case that bore no resemblance to the original Miller design. Following Olen's "go ahead" scrawled on Dodge's memo of December 22nd, 1932, there is not a scrap of paper in the ex-company file that refers to this project of crucial and primary importance. The odds are strong that it was kept top secret, and remained that forever. At this writing, in mid-1995, I don't know of anyone who has ever seen the inside of that case. As already stated, Leo Goossen, having studied photos of it told me that he thought that they might have used three gears, instead of the original two. He heard through the grapevine that it was highly trouble-free. Before this book is published, Christopher Leydon will have dismantled the entire car completely, for the inspection and magnafluxing of each part, and we will know all.* However, on the basis of a pre-WWII Four Wheel Drive Auto Company advertisement from the files of the Ludvigsen Library, London, it seems that transfer was effected by means of two gears and a broad silent chain. Within the final output gear was the central differential, which divided the output between the driveshaft to the rear axle and that to the front. The Four Wheel Drive Auto Company had been perfecting the central differential system since 1910 and transfer by chain for a very long time. It was because it was highly perfect that, in his letter to Miller of October 26, 1931, Olen specified incorporation of "our type of central differential." The history of four-wheel drive in the world might have been totally different had that phrase read, "our type of transfer case with central differential." Mark Dees, in his book *The Miller Dynasty*, notes that the new case had a lever for locking out that differential, if desired. This sounds like a Brisko touch, and it is likely that he was a participant in the project from start to finish.

Illustration 9-5 A 1933 factory photo of much of the powertrain. The V-8 engine is still in place. The carbs are removed and one can look down the inlet ports. The big news is the brand-new transfer case that had been developed for the car by company engineers, with Brisko's race-wise help. (GBC ex FWDCO)

*The car has now been thoroughly gone through. The transfer case contains four herringbone gears, input, two idlers, and output.

Illustration 9-6 Another view of the new transfer case. It had its own dry-sump lube system served by small vertical, cylindrical tank amidships. Here the Miller carbs and Bosch self-starter are in place. (GBC)

Illustration 9-7 The new transfer case from another angle. (RMRC)

Illustration 9-8 The Miller gearbox and clutch with the Four Wheel Drive Auto Company transfer case. (RMRC ex WFMC)

Brisko's qualifying performance at Indianapolis in 1933 triggered two initiatives. One, it spurred Olen to draft this highly informative inter-branch document which sought to explain the why and how of this whole adventure. His failure to speak of Brisko's front-row starting position suggests that everyone knew and was rejoicing over it already.

May 25, 1933

To the FWD Organization:

The Indianapolis race on Decoration Day is the greatest and fastest automobile race in the world. More people come to see this race than any other gathering in the world except Epsom Downs in England. The track is 2 1/2 miles, oblong, built in 1911 of red brick. It is banked at the corners for the first fifty feet from the lower edge of the track at 16 degrees 40 minutes, or approximately 28% grade; the remaining 10 feet at 36 degrees 40 minutes, or approximately 60% grade. The approaches and releases of the turns have a 2 degree bank. The straight stretches are perfectly level. The curves have a radius of approximately 840.4 feet or 7 degrees 20 minutes. Frost heaves have made the track uneven, particularly at the turns, which are on the grade. It is 2 1/2 miles around the track. The track is 50 feet wide on the straight-away, broadening out to 60 feet on the turns, exclusive of concrete safety apron. It has a distance of about 90 feet between the fences. It has a high concrete wall entirely enclosing the track.

THE CENTER DIFFERENTIAL

(X-Ray View of the FWD Five-Speed Transmission)

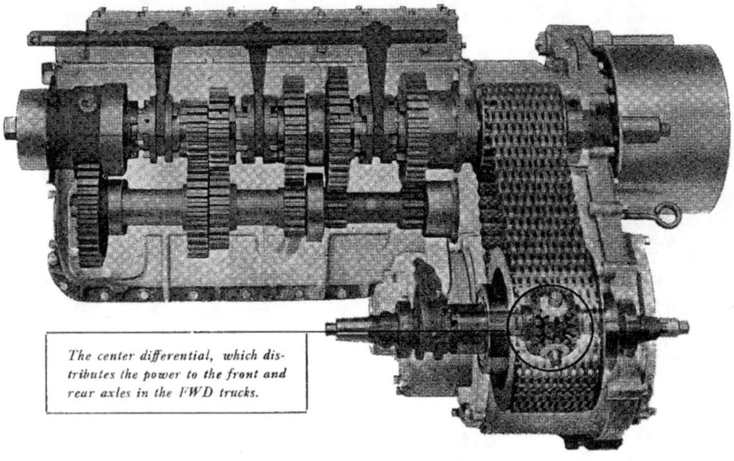

Why a Center Differential is Necessary in a Four-Wheel-Drive Truck

The power in the FWD is distributed to both front and rear axles. The mechanism which divides the power in the FWD is the center differential. Through this device the power is fed from the transmission to both axles. The center differential compensates for the difference in distance traveled by the front and rear wheels and does away with the wear and strain which would otherwise take place. A center differential on a four-wheel-drive truck is just as necessary as the differential in the rear axle of your pleasure car. It serves the same purpose.

When the condition of the road is so unusual that one of the axles loses traction, it is only necessary to engage the center differential lock which is operated from the driver's seat. This gives positive drive to both front and rear axles.

A very effective manner to demonstrate the value of the center differential in the FWD truck is to run the truck in a semi-circle with the center differential locked (this eliminates all differential action), jack up one of the wheels; the strains that are built up by the unequal distance traversed between the front and rear axles will cause the wheels to kick back as much as eight inches with a terrific force. If the truck is operated in the same manner without the center differential locked, the wheels may be jacked up and no strain on the driving mechanism will be shown.

The center differential has been used in FWD trucks since 1910.

THE FOUR WHEEL DRIVE AUTO COMPANY

Clintonville, Wisconsin, U. S. A. **Kitchener, Ontario, Canada**

PRINTED IN U. S. A. FORM N. L. REPRINT 6-14-33—5M

Illustration 9-9 In this 1933 advertisement we have a marvelous view of the internals of the highly distinctive Four Wheel Drive Auto Company transfer case with incorporated central differential.

An average of around 175,000 people are present at the race each year. There is no question but what the greatest judges of high speed in automobiles in the entire world are congregated in one place at this race. Engineers, mechanics and sportsmen come from all over the world to witness the event and to study various angles of the speed problem.

Inspite of the fact that the cars are all technically inspected as to springs, fitness for the race, every precaution being made to prevent cars that are unfit to enter this terrific grind of a continuous speed of over 100 miles per hour for around five to six hours for a total of 500 miles. All of the cars use castor oil for lubrication and about 75 to 80 cars are usually on the track before the race to be tested out and trying to qualify for the race. Only 42 cars showing the highest speed can enter the race, all of which must qualify over the 25 mile run at a speed of 100 or more miles per hour. All the rest are eliminated. The constant running on the track of these different cars with the intense heat they generate makes the castor oil a liquid that freely oozes out of the car and the track becomes saturated with castor oil. The inclines are so slippery that, at times, it is very difficult to walk up by foot without falling or slipping,

With a rear drive car travelling at high speed and entering the curves on the grades at the ends, the skillful driver will bank his front wheels, that is, he turns them up in order to keep on the grade and then in the driver's language "he dynamites his car". That is, he opens the throttle and in this manner keeps up his speed and keeps on the grade. The greatest skill of the driver is tested on the turns on the grades because the car is running around a side hill; the righthand wheels always being higher than the lefthand wheels. With a front wheel drive in reaching this curve on the grade the greatest judgment is in decelerating, otherwise the rear end is apt to cause a spin and both the rear drive and front wheel drivers have to be careful how they bring their cars out of the curve on the grade or again it is apt to produce a skid, a slue or a spin.

All of the drivers who drove the FWD arrived at the same conclusion that they can drive into the curve at full speed, could decelerate or accelerate, did not have to bank the wheels, and, in like manner, could come out of the curve at such speeds as they might desire. They also claim that due to the uneveness of the track there was always a driving wheel on the ground, whereas on both the front drive and the rear drive cars at times the driving wheels left the track. They also found that the control of the car was easier under all conditions.

In the various cars on the track this year the pressure of the tires is 38 pounds. At the end of the 25-mile run the FWD showed a pressure of from 41 to 42 pounds, whereas both the front wheel drives and the rear wheel drives showed a pressure of 44 to 45 pounds, thus further proving that the FWD is easier on tires. Due to the excessive power on each one of these cars there is a marked slippage of tires on the track until the car gets under momentum, which is almost entirely eliminated in the Four Wheel Drive.

The purpose of this racing car was to study high speed under the most difficult conditions and in competition with the greatest drivers in the world. All of the drivers and mechanics have been exceedingly accommodating in giving us the benefit of their knowledge and experience. All of them seem to be very keen on having four wheel drives developed for racing because of the greater safety and the possibilities of greater speed. The saving on gasoline is again showing up on the track, it being about 10%. The saving on tires, so far as we have been able to test, will be at least 20%.

When we think of all the right angle corners, all of the curves and turns, all of the banked roads and that millions of automobiles and trucks are operating over them without any law or regulation as to the leakage of oil and grease, we can gather some conception of how much this is a factor contributing to automobile and truck actions upon the highway. It is the discovery of this fact at the Indianapolis track which has made us concentrate more than ever in trying to perfect the natural advantages of a four wheel drive type of construction, both for propulsion and for braking as well as control for long distance trucking.

We trust that this explanation may be helpful to you in pioneering our truck for long distance work in commercial use.

Very truly yours,

THE FOUR WHEEL DRIVE AUTO COMPANY

Walter A. Olen
President.
W.A.Olen

The other immediate reaction was to launch a new harvest of testimonials, not overlooking such seasoned observers as course marshals. It appears from these documents that several good drivers were given a go at the wheel, in order to have their expert opinions. There is no indication that there ever was monetary recompense for this exchange of privileges. Everyone gained, in terms of experience and public exposure. The harvest began on May 24th with Joe Russo.

May 24, 1933

The Four Wheel Drive Auto Co.,
Clintonville Wisconsin

Gentlemen;

In my opinion, the FWD is physically right. I think that Frank Brisko demonstrated that in his qualification run.

In slowing up for the turns the mass, that is the weight of the car, is shifted to the front and the right front tire, especially, gets the wear. In the case of the front drive, the weight and the thrust on this tire together with the powers increase the wear.

When accelerating out of the turns, the weight of the car is thrown towards the rear. In the case of a rear drive, here again the weight and the power are concentrated; this time in the direction of the right rear.

With a four-wheel drive the actions I have just described aresplit between the front and rear axles. The sudden deceleration for the turns distributes the load equally to four points instead of two, and in this way holds the weight more even. When accelerating out of the turns, four tires take hold instead of two and the stress and wear are spread over the four tires.

It is quite apparent to me that the four-wheel drive is a saver on tires besides having the other advantages of safety at high speed.

Yours truly,

J. E. Russo.

Illustration 9-10 Picture of Russo with Bill Kiser.

Then came the previous 500's winner, Fred Frame.

May 24, 1933

The Four Wheel Drive Auto Co,
Clintonville, Wis.

Gentlemen;

I have been studying the various major improvements made in automotive design and I feel that the FWD as driven by Frank Brisko is demonstrating the advantages of the four-wheel drive. I think that there will be more four-wheel drives in the races hereafter, because of the advantage of more traction. The added traction naturally results in more speed and yet, at the same time there is the saftey factor to go with that speed.

The rear drive has its merits and the front drive has its merits. The four-wheel drive combines the advantages of both. It is a natural developement, for, just as it is only logical to cool an airplane engine with air or a motor-boat engine with water, so it is only logical to utilize all four wheels of a car for traction. The wheels are there; they should be used to the best advantage.

Yours truly,

Fred Frame.

Illustration 9-11 Picture of Frame.

On the 25th, Lou Moore had his say. He was about to finish third in this race and would go to Tripoli with De Paolo and the duplicate four-wheel drive in 1934.

May 25, 1933

The Four wheel Drive Auto Co.,
Clintonville, Wis.

Gentlemen;

The FWD appeals to me because, as a result of its four-wheel drive principle, it should make a better average speed.

It isn't necessary to slow down as gradually for the turns as it is with average car. The traction on all four permits slowing up suddenly without the usual tendency to swing. The four wheel tration pulls the car straight around the curves and it enables the driver to get up speed faster when going into the stretches.

This additional traction which permits a better average speed around the course also serves as a safety factor, for with traction all four wheels there is bound to be a safety advantage all the time, whether the track is dry or slippery.

Yours truly

Lou Moore

Lou Moore

Illustration 9-12 Picture of Moore.

Veteran Zeke Meyer—not a relation to Louis—then doing a superb job with Studebaker Specials (sixth in 1932 and ninth in 1933) spoke. Of course, all of these uniformly typed letters had passed through the same company processing.

May 25, 1933

The Four Wheel Drive Auto Co.,
Clintonville, Wisconsin

Gentlemen;

I have watched a good many cars go bobbing around the corners and swinging back and forth, but when Frankie Brisko went around those curves with that FWD it looked like he was riding on car tracks.

It certainly does handle nice. The necessary traction is there for holding the car in place when slowing up last for the turns. It holds the car in position when going through, and the four wheel traction gives it a quicker get-a-way when coming out.

I think that the four-wheel drive is the coming thing for speed. This principle of drive has demonstrated itself in trucks for getting out of places; now it is demonstrating its advantages for high speed with safety.

Very truly yours,

Zeke Meyer

Zeke Meyer

Illustration 9-13 Picture of Meyer.

Experienced handler of both rear-wheel-drive and front-wheel-drive cars Pete Kreis, had the opportunity to drive this one.

May 27, 1933

The Four Wheel Drive Auto Co.,
Clintonville, Wis.

Gentlemen;

The four-wheel drive race car, in my opinion, is the realization of a racer's dream.

There is, first, the additional traction which is essential to additional speed. The fullest use of the engine power is made to propel the car. With the load and the power distributed to the four wheels, the car has the extra traction which is bound to be helpful in pushing the car at high speed on the stretches.

In the turns this four-wheel traction serves not only as a factor in maintaining speed, but also in maintaining safety, for with four wheels gripping the track, the safety factor has been increased, making it unnecessary to reduce the speed as much as usual to avoid skidding.

When I drove the FWD I was pleasantly surprised at how nicely it handled. It is quite evident that a center differential is necessary for a drive of this type to operate successfully mechanically and from a standpoint of safety.

If I were to own any more race cars I would want them to be four-wheel drives.

Yours truly,

Pete Kreis

Illustration 9-14 Picture of Kreis.

One of the greatest driving and car-building talents of the inter-World War period, Harry Hartz, committed himself strongly. However, his own front-drive machine continued to perform so well that he did not make the predicted conversion.

The Four Wheel Drive Auto Co.,
Clintonville, Wisconsin.

Dear Sirs;

I have been watching with a great deal off interest the performance of the racing cars during the trials this year.

There are a great many types of spring suspensions, both front drive and rear drive, and due to the very high speeds at which the cars are running now, I'm satisfied that the four-wheel drive is the ideal type of car. It is the type of drive that makes the best use of the engine power.

I am planning to build four-wheel drives myself, for I feel that in order to get more speed with safety we have to have traction on all four wheels. It is an outstanding advantage no matter what the condition of the track may be.

You know too, at this high speed the tires get an awful lot of punishment; there is so much power in some in the motors that they're spinning the wheels of some of these two wheel drives. We can't expect the tires to stand all this grief; it would be better to apply the power to the four wheels and save the tires as much as possible.

I'm satisfied, too, that a four-wheel drive can not be a success without a center differential. There would be too much strain on the mechanism of the car and too much danger in the curves.

Very truly yours,

Harry Hartz

Illustration 9-15 Picture of Hartz.

Jean Marcenac, one of the finest racing mechanics ever, came to the U.S.A. with Jean Chassagne's Ballot in 1919, passed to the service of Leon Duray, then Frank Lockhart, eventually becoming chief mechanic of the front-wheel drive Novi operation.

The Four Wheel Drive Auto Co.,
Clintonville, Wisc.

The four- wheel drive principle is right ; no question about it. It's the only way to make the best use of the engine power. By applying the power to all four wheels you're getting away from slippage in the turns and your getting away from a lot of wear on the tires. I wouldn't be surprised if that FWD would go 500 miles on this track on one set of tires, and that is saying something in this racing business.

The cars are going faster this year than ever before and, of course, we want to step up the safety along with the speed. It is only logical that with power on both axles you have the traction which makes the car safer, no matter if the track is dry, wet, or oily.

Truly yours,

Jean Marcenac

Jean Marcenac

Illustration 9-16 Picture of Marcenac.

The real voice of experience was that of Frank Brisko, who had gone much farther and faster in the car than anyone. Still, the points stressed seem to be those that the sponsor wanted to hammer home: stability in cornering; prolonged tire life; the importance of the central differential.

The Four Wheel Drive Auto Co.,
Clintonville, Wis.

Gentlemen;

The FWD has plenty of speed; there is no question about that; the important thing is that I can drive at high speed over the rough, oily track with safety. I go through the curves without a tendency on the part of the car to weave.

Another thing is that this type of drive is easy on the tires. Even though it is a comparatively heavy car, the tire wear and tire heat is reduced because of the power being applied to all four wheels.

A four wheel drive car would not be a success, however, without the differential between the axles. I tried driving with the differential locked, but I had so much trouble in the turns that I fastened the differential control with strap iron so that it couldn't be locked at any time. If there isn't a free-working center differential, there woul be an awful lot of wear on the tires and strain on the mechanism as as well as being hard to handle.

Yours truly,

Frank Brisko

Frank Brisko

The giving of rides and gathering of testimonials took place during the days immediately

Illustration 9-17 Picture of Brisko.

prior to the race. Such topics as the several deaths involved and the consequences of changes in the regulations reduced coverage of the race itself to a minimum in the press. A summary released later by the Company stated, "Held second position for 47 laps. Forced out of race at 118 miles, excessive oil consumption." A chart in the race coverage of *Motor* magazine for June of 1933 showed Brisko/FWD as having been second at the end of 25 miles, third at the end of 50 and then 75. At 100 miles Brisko had regained second place, and at 125 he was shown as lying tenth. The performance of man and car had been splendid, in any case, until about lap 47/117.5 miles, when Brisko pulled into his pit. The rate of loss of his engine lubricant, along with its extreme heat, dictated withdrawal on the spot.

Walter Olen found himself in an agonizingly frustrating situation when he drafted his formal statement for release on the morrow of the moment of truth. The four-wheel drive principle was everything that had been claimed for it. Yet, what Olen chose to describe as the car's "apparent defect" was horrendous to the degree that, in the eyes of an intently watching world, it rendered the four-wheel-drive principle synonymous with failure. His usually facile prose became inept and tangled. What "grades," what "slues," what "company presidents"? Whistling Dixie, he played two cards. One was the "very beautiful, voluntary" letters that had come his way. The other was to dump the whole blame squarely on Harry Miller, who was undergoing the generalized Calvary of his career. For the second time we encounter here the term "subtransmission." Perhaps Olen did not call it "transfer case," as in the past, because he had come to regard the Miller solution as being unworthy of that dignity. The first sentence of his fifth paragraph is deliberately or inadvertently misleading. It leads the reader to think that the car, as it ran at Indianapolis in 1933, had a Miller "sub-transmission." Nowhere is the reader informed of the replacement of the Miller unit by one of four-wheel-drive design. And, the pencilled questioning of the following sentence by an unknown hand is well called for.

5/31/33

The FWD racing car at the Indianapolis track made a very beautiful run for 112 miles. At the end of 100 miles it was in second position. There were no accidents, no skids, no flues. The car was under perfect control at all times. On the grades and on the turns it was outstanding; so much so that even the people in the Grand Stand remarked at the ease with which it operated. The presidents of several large automobile companies arranged to ride in the car in order to study its performance.

There were three men killed at the race and two in the preliminaries, making a total of five deaths. There were several other wrecks and a number of cars had to pull out of the race because of insufficient oil.

The FWD developed trouble with excessive heating of the oil and at the time of leaving the track still had approximately a gallon of oil, but it was heating excessively and leaking oil.

Mr. Frank Brisko, who drove the car, is to be complimented. Not only did he drive it in a admirable manner that brought out distinctly its many fine qualities, but used judgment in bringing the car from the track without harm.

The motor, transmission and sub-transmission in this car was a new design by Harry Miller. The oil circulates from the crankcase into the transmission and back again. A year ago this heating and leakage developed at the end of some 15 to 20 miles. This year it did not show up until at the end of 100 miles. It is a problem which is challenging not only our own engineers and mechanics, but some of the finest experts in the country. It may be necessary to replace the motor and transmission, either in whole or in part, in order to remedy this apparent defect.

All of the features of the Four Wheel Drive for high speed work with safety and control and handling stands out more distinctly than ever before. The company has received some very beautiful letters from guards who were stationed around the track and have observed cars in operation for nine and ten years, all recognizing that the FWD was outstanding in its performance as compared with other rigs. A number of the more prominent drivers in the race voluntarily gave us written statements outlining why the FWD type of a car was superior to others.

As previously announced, we were there primarily to study high speed under dangerous operating conditions and while we are disappointed in the outcome of the race, we are enthusiastic in the very definite outstanding advantages of our car. Our engineering and mechanical ability will be further questioned in remedying the defects in the lubrication of the motor and the transmission, but beyond that, everything is in the car's favor. The car averaged over 114 miles per hour for one hundred miles.

Yours very truly,

FOUR WHEEL DRIVE AUTO CO.

Walter A. Olen

Walter A. Olen
President.
W.A.Olen
ELV

The Company file for 1933 ends with an account rendered by straight-from-the-shoulder Frank Brisko. Whether it was requested or spontaneous doesn't matter; the hard facts and absence of hype are refreshing. The document probably accompanied the handing back of the car to the lessor by the lessee. Early on, we read that "the car" performed perfectly at Indianapolis. Later, the focus of approval is narrowed to "the drive" system. In almost the same breath, Brisko states that he exceeded 119 mph for three laps, which he then bumps to six. The contradiction is unintentional and inconsequential. The car was one of the fastest in the field; possibly the fastest, as some said of it; until the V-8s lower-end bearings were ready to go, or were already gone. For Brisko there was no "engine problem;" one merely swapped the defective one for a type of known reliability.

Milwaukee, Wis.
July 25, 1933.

Dear Mr. Olen:

Writing up a brief statement about the F.W.D. Race Car. I do know that without the center diferential this car would be a failure. I have driven front drive and rear drive race cars and it is impossible to do with them, what you can do with the F. W. D. At Indianapolis this year the car performed perfectly. It was possible to pass cars during the race in any position without the least hint of danger.

The biggest surprize I experienced in my 15 yrs. of raceing, was when I took the car out for the first trial spin in May. When I left I said to my mechanic "I will not try to drive fast as I want to see if our carburation is correct." We went around four laps and I stepped a little harder on the trottle, then pulled in, I figured we had been doing about 110 miles an hour, imagine my surprise when Frame and some of the other boys came up and asked what my hurry was as we turned 3 laps better than 119 miles per hr.

We did have some tough luck at Indianapolis, due to motor failure, but that isn't any reflection on the F.W.D. as the drive worked perfect.

I will again say that the center differential sure does the work. It relieves any undo stress on the driving mechanism, also your load is even at all times

When we qualified for the race we weighed close to 3200 lbs. It would be impossible with this weight and only two wheels pulling to attain the speed the F.W.D. did., six laps during qualifing time we turned better than II9 miles per hr. When we qualified we had a strong tail wind which made us slow down on the back stretch.

At Milwaukee this past June, We got into some trouble. The Car got around on me in one of the turnes. I also made a poor time trial But after analizing just what happened, the car seemed light in the rear and did not balance up, as the gas tank was changed in this car which made the tail light also the transfer case is heavier with an oil tank under the frame, it seems as though the weight is shifted too much forward.

The way it seems and which I believe is correct, the left rear wheel spins, and starts the car into a slide. So I believe that explains the situation in Milwaukee. One more thing I want to explain that on dirt track as no one uses a differential in the rear axel. As when you turn the car wants to tip to the right and lifts the left rear wheel up, I had an open rear axel; that was about 50% of my trouble.

I do not want you to get me wrong, the center differential is O.K., it is the rear axel which should have some kind of a locking device so the left rear wheel will not spin.

Sincerely Yours,

Frank Brisko.

Chapter 10

Wheels Within Wheels

The Deathless Front-Wheel-Drive Mule

What present-day owner David Uihlein, of Germantown, Wisconsin calls "the outboard brake FWD," Leo Goossen identified as "the first Miller front drive." I like to call it "Old No. 1." Many good drivers tried to wring it out when it was new, but its handling was awful and a revised design was made, with inboard brakes and other modifications. It was so much better than No. 1 that Miller's brother-in-law, Dave Lewis, had no trouble qualifying it fifth at Indianapolis in 1925. He finished the five hundred miles only three-tenths of a mile per hour slower than De Paolo's record-breaking average. Old No. 1 didn't even make the race. Peter Shaw, of Intertech Engineering in England, who restored the car for Uihlein, sees application of the outboard brakes having caused the lower springs to wind up. This would bring about a reduction in caster angle and perturbations of steering. In contemporary photos of No. 1, the long, graceful, and slender forks of the De Dion tube are to be

Illustration 10-1 The original Miller outboard-brake front drive, with its gunmetal-blued axles and grille. The long, slender forks at each end of the De Dion tube probably twisted under braking and acceleration, spoiling steering geometry. (GBC ex Packard Motor Car Co.)

Illustration 10-2 Miller 91 front-wheel drive driven by Dave Evans in 1926. The chunky, bent back De Dion tube forks along with inboard brakes were an effort to eliminate the distortions of the past. (GBC ex Leo Goosen)

Illustration 10-3 The outboard-brake car in 1995, as restored by David Uihlein. Its De Dion tube is that of the Miller four-wheel-drive mule, the conversion having been made by Brisko in the 1930s. The forks are a far cry from those of the original axle. Note gunblueing of grille, axle. (GBC ex David Uihlein)

admired. Also, one can readily imagine them distorting under acceleration and braking. This may explain why the front brakes of future models were situated inboard and the kingpin forks were made shorter, more stubby, and were curved far more abruptly to the rear. This trend culminated in the relatively ungraceful and chunky De Dion tube of the Miller four-wheel drives.

No. 1, then, was the mule* of Miller front-wheel drive, a lame duckling that, nevertheless, had an almost uncannily long and active career. As we have seen, Miller immediately shunted the mule to Packard. With the same alacrity, B.M. Ikert, one of the best technical writers of the day, filed a story which appeared in the July 30, 1925 issue of the magazine *Motor Age*. Its lead photo was of the second-place car streaking past a crowded grandstand, and its title was "Driving All Wheels Of A Car." Evans's car, Ikert wrote, "upset a good many theories about such cars being unmanageable; tricky to steer, and so on. It is true the power in this car was not applied to all four wheels, but during the race it was clearly demonstrated that applying the power to the front wheels need not interfere with steering." And that was the most basic key to all-wheel drive.

Packard engineers learned all that they could from the mule and, after a year or so, it was sold to one Stanley L. Reed. Reed reworked the car's engine to bring it within the 91 cu. in. limit; and, with one Bruce Miller at the controls, tried to qualify the car for the 1927 Indianapolis 500. It did not make that race, but qualified sixteenth in 1928 and thirty-first in 1929, lasting 132 and 50 laps, respectively. Reed then sold the machine to Hank Maley, who later would provide the Duesenberg chassis to receive the Scully 4C255 engine. Maley punched the old front-drive's engine out to 138 cu. in. Dave Evans qualified the car in thirty-third place for the 1930 Indianapolis 500, but finished a fine sixth. Then, who should buy the machine but our friend Frank Brisko. He managed to stretch the engine's displacement to 151 cu. in. He qualified the car for the 1931 Indianapolis 500 in twenty-seventh place and stroked along until a steering arm broke on the 138th lap. In 1932, Frank tried again, this time qualifying thirteenth, at a not bad 111.1 mph, only to have his clutch pack up on lap 61. Then, as we know, he secured the lease of the Four Wheel Drive Auto Company four-wheel-drive car. He entrusted Old No. 1 to Wesley Crawford to drive in the 1933 Indianapolis race. Crawford qualified twenty-sixth and on lap 147 a front half-axle seems to have sheared and the car crashed, wiping out its front end, but without serious injury to driver or mechanic. Brisko put the wreck in storage and, as we shall see, eventually replaced its front end with that of the four-wheel-drive mule. That these two prototypes—the Miller front-wheel drive prototype and the four-wheel drive one—born almost a decade apart, ultimately should be fused into a single entity gives one something to reflect upon. It makes one think of Miller's words, "I don't do these things. I have a control that tells me what to do. I count on it."

*A "mule" is simply a car used for development.

Illustration 10-4 The original outboard-brake front wheel drive, recuperated by Brisko and campaigned by him for years. Note the suspension, identical to that of the Miller four-wheel drives of 1932. (GBC ex IMS)

The Four-Wheel Drive Land Speed Record Car

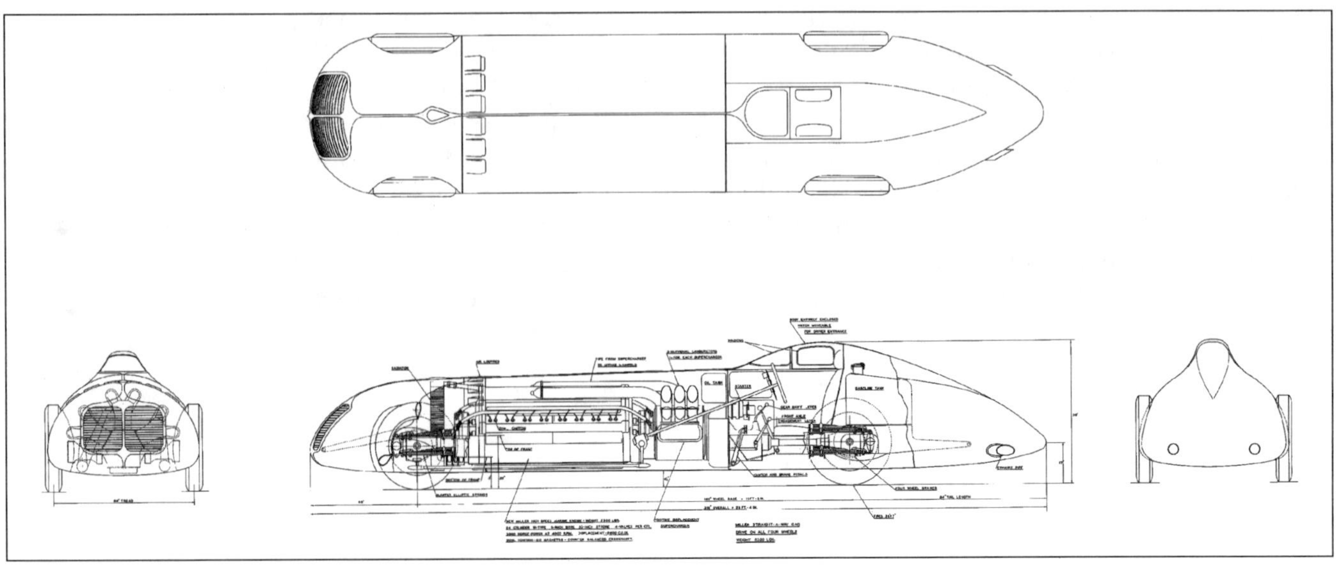

Illustration 10-5 Plans for the Miller four-wheel drive Land Speed Record car. Note quarter-elliptic suspension front and rear, as on the 1931 series racing cars and four-wheel drives. The huge Roots blower, between the W-24-cylinder engine and the bell-housing follows the principle pioneered on the Burnden super road car. This drawing presumably was made by Myron Stevens. (FAUC)

Illustration 10-6 Barney Oldfield, long retired, was eager to drive the projected four-wheel drive Land Speed Record car at better than 300 mph. (GBC)

Illustration 10-7A As far as the project ever got: the same 1/8 scale model decades later, photographed by Strother MacMinn. (GBC)

Illustration 10-7B Another picture of the 1/8 scale model, photographed by Strother MacMinn. (GBC)

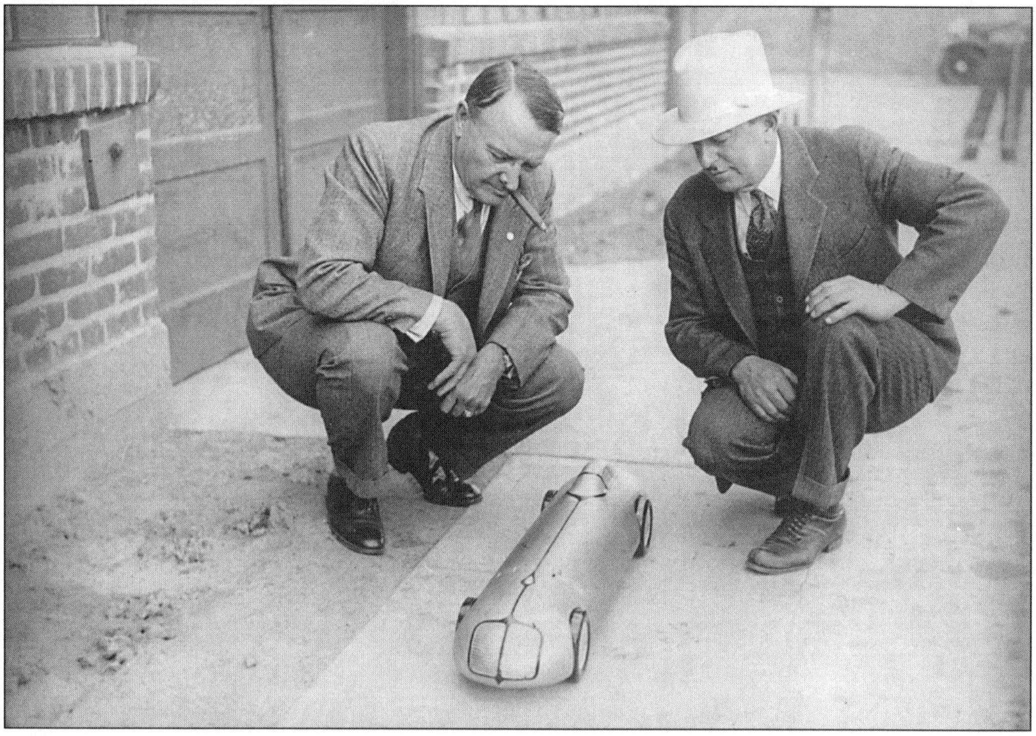

Illustration 10-8 America's most legendary kings of speed—Barney Oldfield and Harry Miller—contemplating the heir to the world's speed title. (GBC ex FAUC)

Mark Dees writes, "The ax fell on July 8, 1933, when his outside pattern-makers and foundry filed petition in federal court which placed Miller in involuntary bankruptcy." A U.S. Marshal padlocked the doors of the plant, but many people knew the contents of that Ali Baba's cave. Many treasures disappeared at night, through other openings in the building, long before the formal auction on October 11th. Among the dreams that evaporated when the ax fell was the Land Speed Record project which Harry Miller had pitched to Walter Olen in his letter of July 3, 1931 (see Chapter 1, page 4). If not at the outset, the project soon evolved into a joint-celebrity undertaking: Miller would design and build the car and Barney Oldfield would drive it, going after the 253 mph land speed record then held by Sir Malcolm Campbell. A celebrity-to-be, Preston Tucker, was hustling in the wings. The conception was complete and all that was needed was someone to bankroll it. Harry Miller kept trying until well into 1933.

The car, theoretically capable of around 360 mph, would get its thrust from a 2600 cu. in. W-24. It would be capable of delivering around 3000 bhp at 4000 rpm. Its wheelbase would be 186 in./15 ft. 6 in. and overall length 316 in./26 ft. 4 in. It would weight 6500 lb., about 1500 less than Sir Malcolm's big, finned Bluebird.

"The fin will be unnecessary," said Oldfield in an article of unknown date in *Popular Mechanics*, "because the Miller Special has four wheel drive, whereas the Bluebird drives from the rear wheels alone." He justified this effort saying, "When we have the superhighways built that are now planned, we will tour at 100 miles an hour; perhaps at greater speeds; so we must continue to push up the speed record with special cars so that tomorrow all the world may attain tremendous speed with safety."

According to Oldfield, the body had been wind-tunnel tested in scale model form. Who designed the body and what its aerodynamic characteristics were seems not to be known. One unusual feature of the machine was the location of a Roots blower, about the size of a fifty-five-gallon drum, between the engine and clutch housing. This arrangement, of course, was used on the Burden car. The blower of the Land Speed Record car was to have three carburetors, each feeding an inlet manifold to a separate bank of eight cylinders. Another notable point is that the structural layout of the Land Speed Record chassis was identical to that which Miller had adopted for his speedway cars: deep channel-section frame side members and over-and-under quarter-elliptic springs at both front and rear. All of this is directly traceable to the No. 1 front-wheel drive. The only drawing of the powertrain is a side elevation which does not show a transfer case or other details of how the power was to be transmitted to the front wheels. Drive to the rear appeared to be conventional.

Bunny, Scully, and Pete

The bankruptcy sale on October 11, 1933, must have attracted a memorable turnout of "Who Was Who" in the local auto racing world, plus not a few pilgrims from afar. It is one of my great regrets that, in talking with old-timers who had been there, I never asked for a description of what that dramatic event had been like. There were always more urgent things to pin down. I can't believe that Harry Miller was present. I don't see how he could have survived that trauma. However, people did tell me that, throughout that whole experience of being stripped of practically everything, one could not tell by looking at Harry Miller, or by his behavior, that he had really been touched by it all. The people at the sale who concern us in the four-wheel-drive context are boat-racer Dick Loynes, plus Frank Scully and his partner, Bunny Phillips. And possibly Frank Brisko.

Illustration 10-9 The man to beat in both national and international 155 cu. in. hydroplane racing in the late 1920s was Dick Loynes. At the awful Miller bankruptcy sale he picked up all the patterns, drawings, several engines, and many parts for a song. (GBC)

Illustration 10-10
Bunny Phillips.

Frank Scully, out of Chicago, and Overton A. "Bunny" Phillips, essentially an Angelino, met when the latter advertised in a local newspaper that he had a Bugatti for sale. It was an early straight eight. He had recently lucked into a practically new type 35B Bugatti and wanted to go road racing with it in Europe. He had other cars and was unloading everything to finance his trip. Scully had already been to Europe, had seen a lot of racing, and had acquired a passion for Bugattis. The upshot of all this was that Frank talked Bunny into establishing the first "official" Bugatti service center in the U.S.A., right there in Hollywood. They didn't have to handle only Bugattis; they could handle everything that was interesting and exciting. Bunny could use the place as a base for his own racing activities, but locally. So, some time in 1930, Bugatti Service (factory authorized parts) opened its doors at 1222 North Western Avenue. There were about six cars of that marque in Southern California.

Bunny and Frank survived, in spite of this and the depression, and got in a nice amount of racing with Bunny's Type 35B. Then, in 1932, at an outlaw 100-miler in Oakland, California, Bunny's engine blew itself to bits. It was nearly irreparable. What to do? It was at this point that Bunny, at Bugatti Service, installed a flat-head Ford V-8 in the Chancellor Miller, in exchange for its magnificent but abused engine, which he planned to restore to health and install in the Type 35B. It was a swell idea, but the engine's cylinder banks were splayed at 90 degrees and would interfere with the Type 35B's exquisite steering. He eventually sold that jewel of an engine to Dick Loynes, after which it vanished.

These men were partners, and in the winter of 1932–1933, Scully bought a Miller 4C, which was one of the very first of the Offy 255s. This bomb was intended to go into the Type 35B. However, retired driver, but evergreen hustler, Pete De Paolo connected with Scully ("got to him," Bunny said) with a much more imaginative plan. He knew a man in Indianapolis named Hank Maley, whom we have already met in connection with Old No. 1. Maley had the chassis of a 155 Duesenberg that was idle. Pete knew a great young driver, Lou Moore, who had just qualified fastest at Indianapolis. "With your engine in that car, and with Lou to drive, we could clean up," Pete told Scully. So Bunny got to install the howling Offy-to-be in the Maley/Scully entry for the 1933 Indianapolis 500. Result: Pete had been right. Lou Moore qualified the car fourth and finished third.* The group had won a substantial chunk of the world's richest motor racing purse, rivaled only by the Grand Prix of Tripoli. They went seriously into the business, as California Racers, Inc.

From May 30th, 1933, we move to the bankruptcy sale on October 11th. There, Dick Loynes bought, for an incredible $25, the accumulated mass of Miller drawings. They filled three mattress cartons. And Frank Scully bought the four-wheel-drive mule, paying $1400 or $1600, depending on the reference source.

Perhaps the car was not worth a great deal more at the time. It was acquired for cannibalistic motives, probably envisaged by De Paolo and Frank Scully, and gone along with by Bunny. The gem of the deal was the engine which, with its cylinder banks splayed at an offbeat 45 degrees, would slip neatly into the space available in Bunny's languishing Type 35B Bugatti. Pete had plans for the rest of the carcass, too. But, before any of this could transpire, he had a plan that would give the four-wheel-drive mule one last, glorious fling and, at the same time, would so fulfill Frank Scully's yearning to mix it up with the big-time Europeans on their own and neighboring turf. Beyond that, it would pull Pete out of the mothballs of retirement and permit him to go out in beauty as an international Grand Prix heavyweight. As we shall see, the combinazione included even more than that.

The Miller-Marmon Connection

Miller lost his home, everything. Fortunately, his beautiful and devoted wife had providently put away an emergency fund with which they managed to survive. One of the L-29 Cords must have been registered in her name or in son Ted's. Financially, Harry Miller was thoroughly discredited in California, the place was full of wrenching reminders, and everything pointed to making a fresh start elsewhere. Although Preston Tucker was, or soon would be, transportation manager of a large brewery in Detroit, in charge of the truck fleet, he always had plenty of other irons in the fire. "What sort of a man was he?," I asked Ted. "Well," Ted replied, "he wanted to be a con man, but he didn't know how."

The story of the exodus from Southern California was recounted by Ted to his listeners as follows: "Tucker somehow arranged for my father to build 50 armored cars for what was then the Persian government, and to do it in the Marmon plant in Indianapolis. I drove my parents all the way back there, in the Cord, arriving just a few days before Christmas. We

*His engine is said to have been the first Miller 255 to go the full distance in an Indianapolis 500 race.

Illustration 10-11
Preston Tucker. (IMS)

stayed at the Hotel Lincoln." Ted then told me a vague story of a couple of months' delay while a new corporation was established as a legal foundation for this new activity. About a week after that, he said, the Wall Street operator who was brokering the deal committed suicide, whereupon the whole thing fell apart. The family moved to Michigan and Tucker put Ted to work driving a beer truck and collecting proxies for Tucker to use in a takeover attempt on his employer's business.

Mark Dees received a variation on this theme, which enabled him to write in his book, *The Miller Dynasty*: "He [Harry Miller] loaned Preston Tucker the use of his name to support an abortive effort to resuscitate and control the failing Marmon Motor Car Company in Indianapolis. In connection with this, Ted Miller recalls that his father designed a four-wheel-drive armored car for Marmon-Herrington, an outfit which was something of a successor to the previous firm." The reality was and is otherwise.

Meet Arthur Herrington

All too little is known concerning the career of this giant of automotive sport nationally and of the motor vehicle industry worldwide. The best source, by far, is the award-winning 1985 book by George and Stacey Hanley, *The Marmon Heritage*. The first half of that tome is devoted to the great old Nordyke & Marmon Co., which metamorphosed into Marmon-

Illustration 10-12 Col. A. Herrington.

Herrington during the very early thirties. The second part of the Hanley book deals with that firm, which flourishes today as the cornerstone of the megacorporation known as The Marmon Group. Art Herrington quarried and laid its cornerstone.

Arthur William Snyder Herrington, well known as "AWH" and "the Colonel," was born in Coddenham, England, on March 30th, 1890. When he was five years old his family moved to New Jersey. He eventually earned a degree in mechanical engineering from the Stevens Institute of Technology and, liking motorcycles, went to work for Harley Davidson. With American entry into World War I, he became a dispatch rider in the Motor Transport Corps of the U.S. Army overseas, with the rank of Lieutenant. He was well placed to observe military vehicles of every type working under real war conditions. After the war he became based in Washington, D.C., as a consultant on off-road vehicles to the Army and Marine Corps. He was tall, with an imposingly lofty brow and military bearing, which gave him his "Colonel" nickname, although he never held that rank.

An eternal interest in motor racing both as sport and as technological discipline drew Herrington to the Indianapolis 500, beginning with its first installment, in 1911. That trial of course was won by Ray Harroun, in a Marmon. Art's involvement increased to the point that he was Jimmy Murphy's pit manager in 1922, when Murphy won the Indianapolis 500 with a Miller engine in his Duesenberg. When Pete De Paolo won in 1925, in a Duesenberg, Art was in charge of his pit. He then was appointed to the Speedway's Technical Committee, eventually becoming its Chairman. He was the last Chairman of the A.A.A. Contest Board. In 1954, when that Board was dissolved following the great disaster at Le Mans, it was Art Herrington who, with Art Pillsbury, created the United States Auto Club, U.S.A.C., to replace it.

Careerwise, off-road transport remained Art Herrington's specialty. An eyewitness to front-wheel drive's first great manifestation, at Indianapolis in 1925, he evaluated it in those terms. Lindbergh's transatlantic solo flight in May of 1927 triggered the flowering of commercial aviation which, in turn, created new ground transport needs. And introduction of the Miller-inspired and derived L-29 Cord in August 1929 marked the arrival of front-wheel drive as a practical industrial reality.

Another milestone in Herrington's career was his attendance at the Military Pageant and Exposition that took place in June 1929 at the Army's Quartermaster Depot near Baltimore. In the words of the Hanleys, "That event provided commercial truck builders an opportunity to compare the performance of their vehicles with military-designed prototype trucks on the same test course." It provided observers with the same opportunity, which was important in leading Herrington to visualize a modern range of all-wheel drive vehicles. One was a refueling truck for the country's burgeoning new airports; a type with good off-runway capability. From his racing experience he knew that a key to the problem would be constant-velocity universal joints in the hubs of the steered wheels of the vehicles to come. He went to work on the design of a CV joint of his own. When finished, it bore a striking resemblance to the double-cardan solution that Leo Goossen had devised for Cornelius van Ranst to use in the L-29 Cord. Herrington applied for a patent on his device on May 15th, 1930.

With this patent pending and his mind teeming with ideas for its application to the all-wheel-drive future, Herrington sought an organization to develop and exploit them. Right there in Indianapolis he knew industrialist Frederick E. Moskovics, who also was an eminence on the administrative level of big-time American auto racing. Moskovics had just finished a stint as president of the Stutz Motorcar Co. there. Prior to that he had been a top executive at Nordyke & Marmon, also of Indianapolis. He still was one of the few stockholders who was not a member of the Marmon family, and he knew the family well, including Walter C. Marmon, Chairman of the Board. In the grip of the Depression, the old firm was sinking fast, its splendid factory facilities with it.

Herrington seems already to have had in hand an Army contract for thirty-three four-wheel drive trucks. With this as a start, the Marmon-Herrington Corporation was constituted on March 10, 1931, with Frederick E. Moskovics as Chairman and major investor, and Art Herrington as Vice President and chief engineer. The old firm and the new were legally interlocked. Within about two months the prototype of the new Army vehicle was running and a great new manufacturer of all-wheel drive vehicles was on the road to far-flung, clamorous successes. Those extended to the Middle East in a big way in 1932, and in the winter of 1933, Art Herrington was given great personal authority over the design and construction of equipment for the mechanization of the Imperial Persian Army.

This is where we came in, to the garbled tune of the Preston Tucker-Marmon-Persian government operation of the winter of 1933. It is eminently possible that Tucker promoted Miller's consulting services to Herrington at that time. I once saw a copy of the Articles of Incorporation pertaining to this Tucker-Miller exercise. I recall being struck by the fact that Miller's name did not appear in them as an officer; only Tucker's. This may be why Harry Miller's son, Ted, chose to say that his father had "lent his name to Tucker" at that time. It is equally possible that, after a certain amount of contact between the three men, Herrington decided that what those others had to offer was without interest, and sent them on their way.

The Herrington patent, which, in a way, had been the foundation stone of the Marmon-Herrington empire, was not granted until November 20th, 1934. As often happens, the drawings that accompanied the application were impressive and complex. But the only real novelty discerned by the examiners was in the sort of linkage joining the two cardan joints. Herrington wasted no time there, moving on to new principles.

The Central Differential

Art Herrington was a harsh critic of existing all-wheel-drive practice. That it was archaic in his esteem was the rock upon which his new empire was founded. He expunged the central differential from his product line and, on any podium, scoffed at its continuing use by others. In 1995 I asked Marmon-Herrington general sales manager William Riley if this bit of company design philosophy still was valid. His reply was clear and precise:

"Not using a transfer-case differential is still our standard. If differential action is required, there is no reason to be in all-wheel drive. If all-wheel drive is required, the driver normally would lock out the transfer-case differential anyway."

It is interesting to weigh this conviction relative to those expressed on the same subject by spokesmen for Clintonville and by Bill Milliken.

Chapter 11

Back to the Four-Banger Future

The Four Wheel Drive Auto Company correspondence file ground to a halt in 1933 with Brisko's letter of July 25, in which he pronounced the central differential "O.K." but emphasized the necessity for a means of locking out the differential in the rear axle. Nothing was preserved in writing concerning the propulsion unit, the handsome, narrow-angle V-8 that was disposed systematically to wipe out its lower-end bearings.

The file for 1934 opens with a response from Fred Offenhauser, now the head of his own company, to an inquiry from the Four Wheel Drive Auto Company, and no doubt from Dodge, concerning crankshafts. Obviously, that of the V-8 had to be replaced. Fred's letters have the air of having been typed by him personally. Since work was so scarce, he certainly had the time to devote to that.

January 14, 1934

Four Wheel Drive Auto Co.,
Clintonville Wisconsin Attention- Mr. H.B. Dodge

Dear Sir:

This acknowledges receipt of your letter dated Jan 13th referring to the matter of Crank Shafts and the material used in them. There are more shafts made of the Chrome Nickel Moly Steel than Nitrolloy steel but the reason for this is that they cost less and also if a new shaft is required in a hurry it can be produced quicker than a Nitralloy shaft on account of the long time required for Nitriding.

The Nitralloy shafts are very satisfactory and there has been only one that broke and that was due to the shaft being made to light and I don't think there is any question but what the Nitralloy Steel makes the best Shaft.

In regards to hardening the Nitralloy shafts we harden them all over except the threads on the Gear end and in the Fly Wheel Bolt Holes it making the shaft much more stiffer, if we were to make a shaft out of Chrome Nickle Moly Steel we would have to get the Billet of Steel from Chicago as there is none on the coast and that would take two weeks.

The price of a Chrome Nickle Moly shaft is Two hundred and fifty dollars ($250.00) and it would be heat treated after rough machining and would have a shore hardness of 45 and would take between four and five weeks for delivery on account of getting the steel from the East.

The terms would have to stay the same as quoted in our letter of Jan 5th, as I only have a small plant and can't carry on this special work any other way. I might suggest that you consult Mr. Frank Brisko as to the quality of work we do.

Yours truly,

Offenhauser Engineering Co.

Fred Offenhauser

Fred Offenhauser

But no. The following day Offenhauser received an order from Brisko, dated January 15th. It was for a set of connecting rods for a Miller 255 four-banger. From here onward the Company file is not its old, ultracomplete self, although it is very rich for this year. From Fred's letter to Brisko dated the 17th we begin to piece together what had been going on since last July. Brisko would seem to have persuaded the company to replace the V-8 engine with a 255—a 255 that he, Brisko, would build, right there in the Clintonville factory. This, according to what we have always believed, would mean ordering the necessary castings and a set of blueprints from Dick Loynes, who had obtained the patterns and drawings for a pittance at the bankruptcy sale. In this letter, Fred asked Frank to let him have a set of copies of the 255 drawings that had been obtained from Loynes by Brisko. Business was looking up a bit, and we see that Lou Moore was busy in his involvement with Bunny, Scully, and Pete.

January 17, 1934

Frank Brisko
Clintonville Wisconsin

Dear Frank:

Received your letter of Jan 15th, ordering the four 255 CuIn rods which I had in stock almost completed and which I will finish and have ready to ship to you in 10days and the price will be $45.00 each as I quoted you when you were out here ordinarly the price of 255 rods are $50.00 each and the 220 rods $45.00 each.

Louis Meyers engine is pretty well along I just got finished with all the parts we made for him. At the present time we are building a Transmission and Clutch Unit for Lou Moore and also for Deacon Litz besides a 220 Cu In engine for Russ Garnant as well as quite a few other parts.

Frank if you can send me those 255 photographed prints you spoke about I sure would appreciate it.

Thanking you for this order, and if there is anything I can do for you out here all you got to do is ask me. Hoping you are feeling fine.

With best regards,

Offenhauser Engineering Co.

Fred Offenhauser

Fred Offenhauser

P.S. The weather is fine again just like when you got here but had a Hell of a Rain Storm New Years day.

Swapping the 255 for the V-8 was facilitated by the fact that both engines were of identical length.

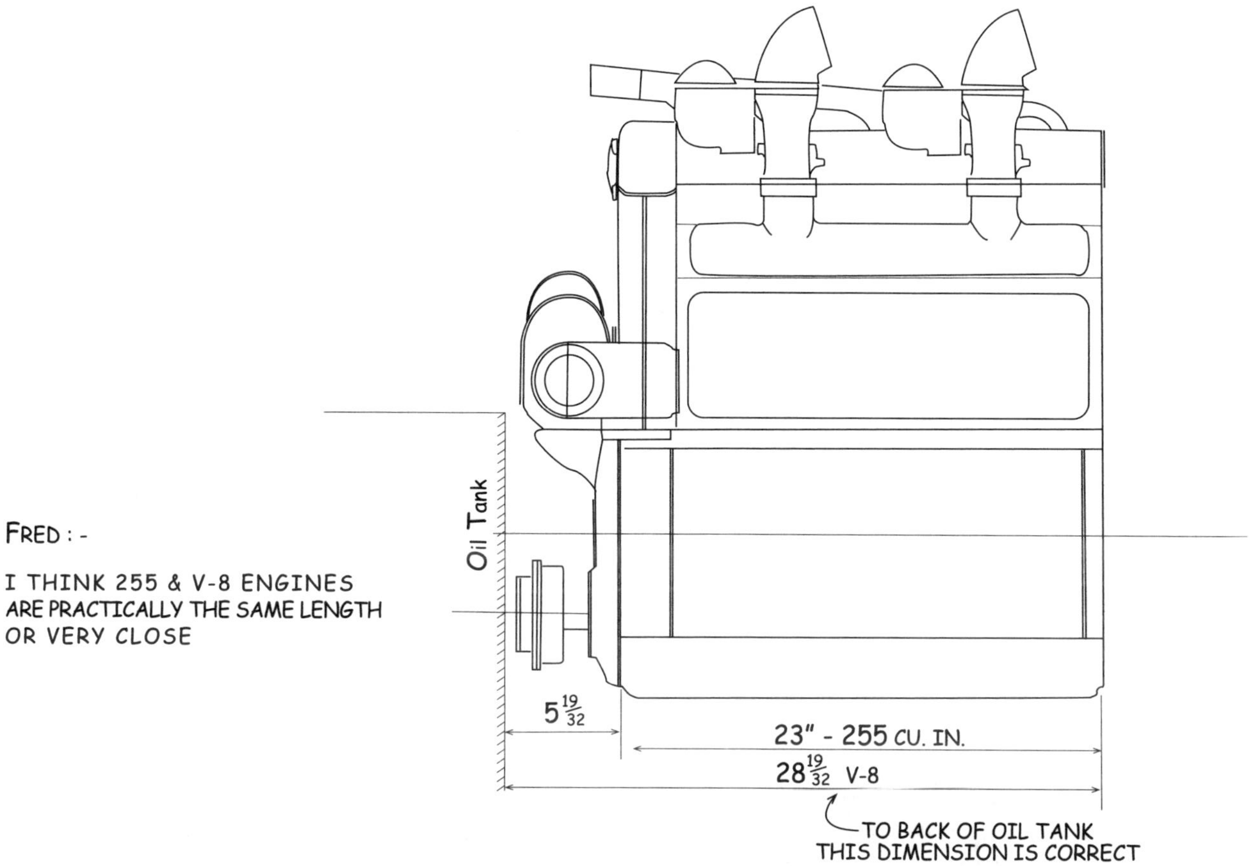

Illustration 11-1 255 cu. in. Miller/Offenhauser engine.

In Fred's letter to Brisko dated February 1, 1934, we learn how Fred got the drawings for what soon would become the Offy 255 and, in 1937, the Offy 270.

February 1, 1934

Mr. Frank Brisko

Clintonville Wisconsin

Dear Frank:

Received your letter of Jan 26th, and enclosed you will find the 2 Blue Prints that you wont. I also wish to thank you for those Photographic copies of the 255 prints.

I am shipping the rods to-day. With best regards I remain,

Yours very truly,
Offenhauser Engineering Co.

Fred Offenhauser

Fred Offenhauser

The first Four Wheel Drive Auto Company document of 1934 that we have is a letter from Dodge to Olen dated February 22nd. Instead of ordering the crankshaft for their 255 from arch-master-machinist Offenhauser, who knew all of the specific pitfalls, they no doubt had adequate reason to give the job to Merz Engineering in Indianapolis. It is nice to note that this company was formed in late 1927 around the machine tools of the recently bankrupt H.C.S. Motor Co. of Harry C. Stutz. In 1939 Merz became part of the all-wheel drive empire of Marmon-Herrington. At the bottom of this letter are the OK and initials of Walter Olen.

February 22, 1934

Mr. Olen:

We have been disappointed in obtaining a crankshaft for our racing car and now have a wire to the effect that the second shaft has cracked during heat-treatment and we cannot obtain a finished shaft before March 27.

The Merz Engineering Company are having unusual tough luck in making this shaft as it is very seldom a flaw occurs. However, I fear that they may experience a similar failure on the next shaft and, therefore, recommend that we authorize them to proceed with all haste and at the same time we order a piece of steel and machine a shaft in our own shop so that should the Merz Company experience another failure we will then have a shaft of our own manufacture to fall back on.

As time is so short, I believe we should not take a chance on simply providing one shaft and in the event both shafts turn out to be perfect I believe we can dispose of one shaft.

The cost of shaft if made by us should not exceed $275.00.

Yours very truly,

H.B.Dodge

H.B.Dodge
Chief Engineer
MIM
(handwritten note): OK WAO

This is the wire from Merz that triggered the foregoing message from Dodge to Olen.

```
GU 68 DL XU=AU INDIANAPOLIS IND FEB 22 1934 207P

H B DODGE,CHIEF ENGINEER=
FOUR WHEEL DRIVE AUTO CO CLINTONVILLE WIS=

REGRET INFORM YOU SECOND CRANKSHAFT REVEALS FLAW RUNNING LONGITU-
DINAL ON MAIL BEARING ALMOST SAME LOCATION AS FIRST CRANKSHAFT
HOWEVER SHAFT HAS BEEN HEATTREATED AND FINISH MACHINED WITHIN FEW
THOUSANDTHS BEFORE DEFECT SHOWED STOP STEEL FURNISHED BY NATIONAL
FORGE AND ORDINANCE SAME SPECIFICATIONS AS USED BY WRIGHT FIELD
AND ALLISON ENGINEERING STOP WILL YOU STILL HAVE TIME ALLOW US
FURNISHED THIRD SHAFT DELIVERY DATE MARCH TWENTY SEVENTH WIRE
ANSWER=

MERZ ENGINEERING CO.
232P.
```

Hand-dated "March 16" and marked "File with Race Car" is the following fascinating announcement from the Automotive Gear Works of Richmond, Indiana. During this period the American oval-track race car industry depended for its final-drive gears mainly on those

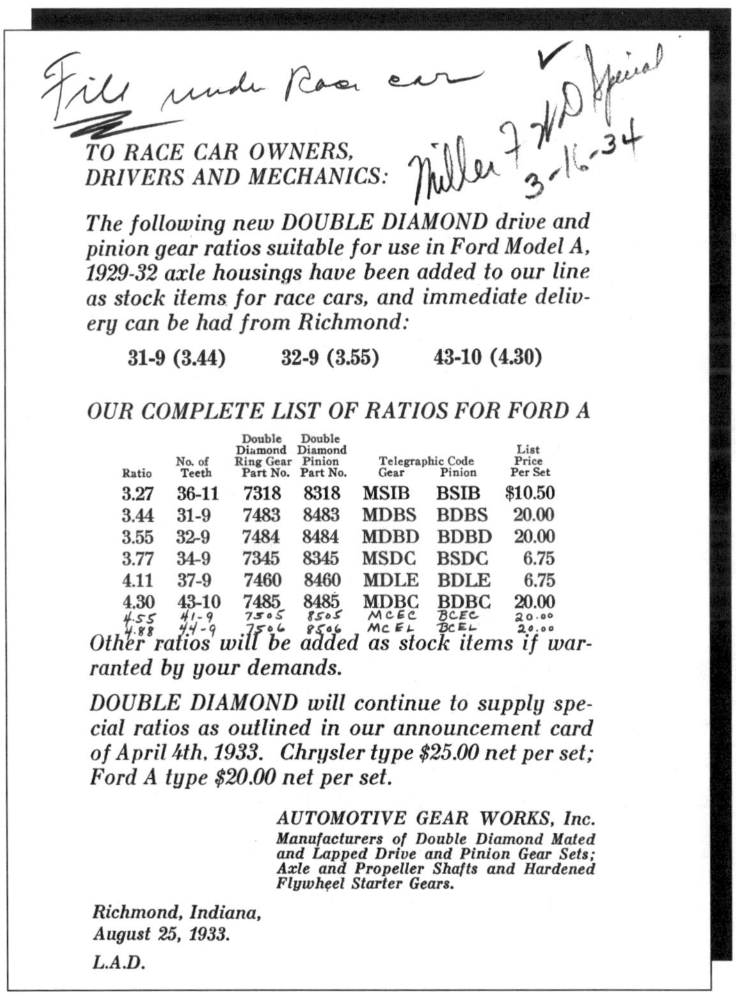

File under Race car ✓ Miller 7 WD Special 3-16-34

TO RACE CAR OWNERS,
DRIVERS AND MECHANICS:

The following new DOUBLE DIAMOND drive and pinion gear ratios suitable for use in Ford Model A, 1929-32 axle housings have been added to our line as stock items for race cars, and immediate delivery can be had from Richmond:

31-9 (3.44) 32-9 (3.55) 43-10 (4.30)

OUR COMPLETE LIST OF RATIOS FOR FORD A

Ratio	No. of Teeth	Double Diamond Ring Gear Part No.	Double Diamond Pinion Part No.	Telegraphic Code Gear	Telegraphic Code Pinion	List Price Per Set
3.27	36-11	7318	8318	MSIB	BSIB	$10.50
3.44	31-9	7483	8483	MDBS	BDBS	20.00
3.55	32-9	7484	8484	MDBD	BDBD	20.00
3.77	34-9	7345	8345	MSDC	BSDC	6.75
4.11	37-9	7460	8460	MDLE	BDLE	6.75
4.30	43-10	7485	8485	MDBC	BDBC	20.00
4.55	41-9	7505	8505	MCEC	BCEC	20.00
4.88	44-9	7506	8506	MCEL	BCEL	20.00

Other ratios will be added as stock items if warranted by your demands.

DOUBLE DIAMOND will continue to supply special ratios as outlined in our announcement card of April 4th, 1933. Chrysler type $25.00 net per set; Ford A type $20.00 net per set.

AUTOMOTIVE GEAR WORKS, Inc.
Manufacturers of Double Diamond Mated and Lapped Drive and Pinion Gear Sets; Axle and Propeller Shafts and Hardened Flywheel Starter Gears.

Richmond, Indiana,
August 25, 1933.
L.A.D.

Illustration 11-1
Announcement from Automotive Gear Works.

for the Model A Ford. The entire rear-axle assembly was a classic component for many builders. This is a testimony to the margin of safety and to the metallurgical excellence of Ford products. The Ford A developed 40 bhp, but its rear-end gears did not fear 300. Model T Ford rear-axle half-shafts were the choice of all the great circuses for tent stakes. Their steel remained unfazed by the constant pummeling of heavy sledge hammers.

We do not know on whose dynamometer the power and torque curves for the Clintonville 4C255 were determined.

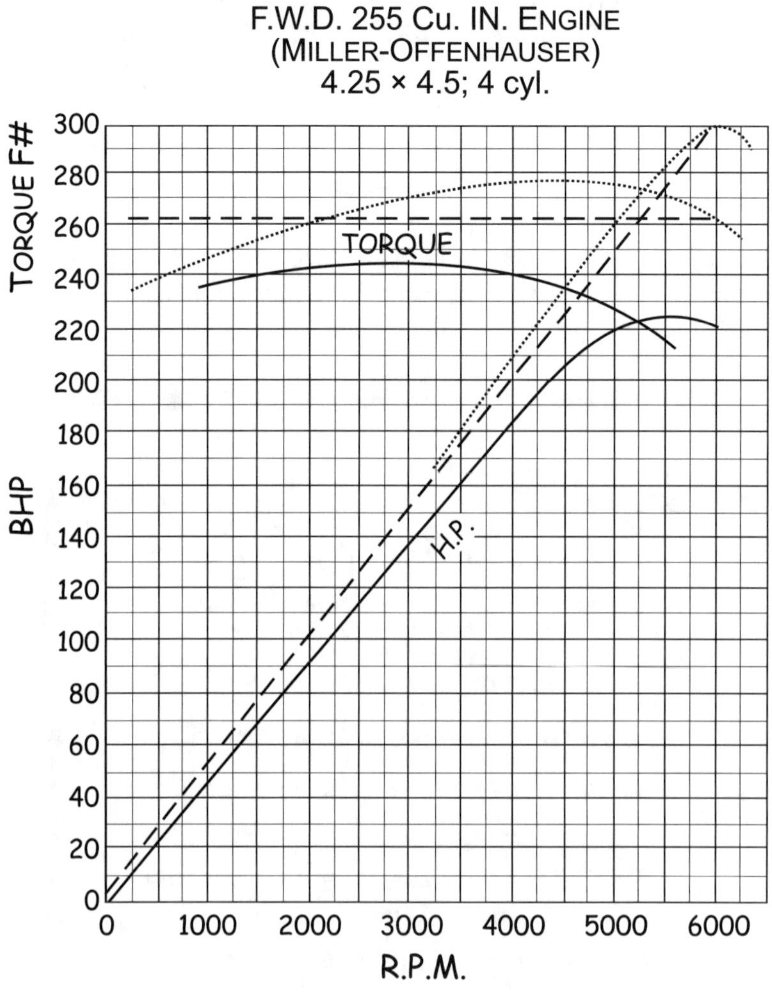

Illustration 11-2 Horsepower and torque curves.

And these presumably are calculated speeds of the car with 255 engine in various gears. The gearbox ratios shown are probably those of the four-speed Ford truck transmission, which was widely used in American race cars during this period.

The crankshaft problem was somehow overcome, the 255 installed, and the car sent off to the Indianapolis Speedway, in Brisko's care. He had not forgotten about locking out the rear-axle differential, and this experiment also received Olen's OK.

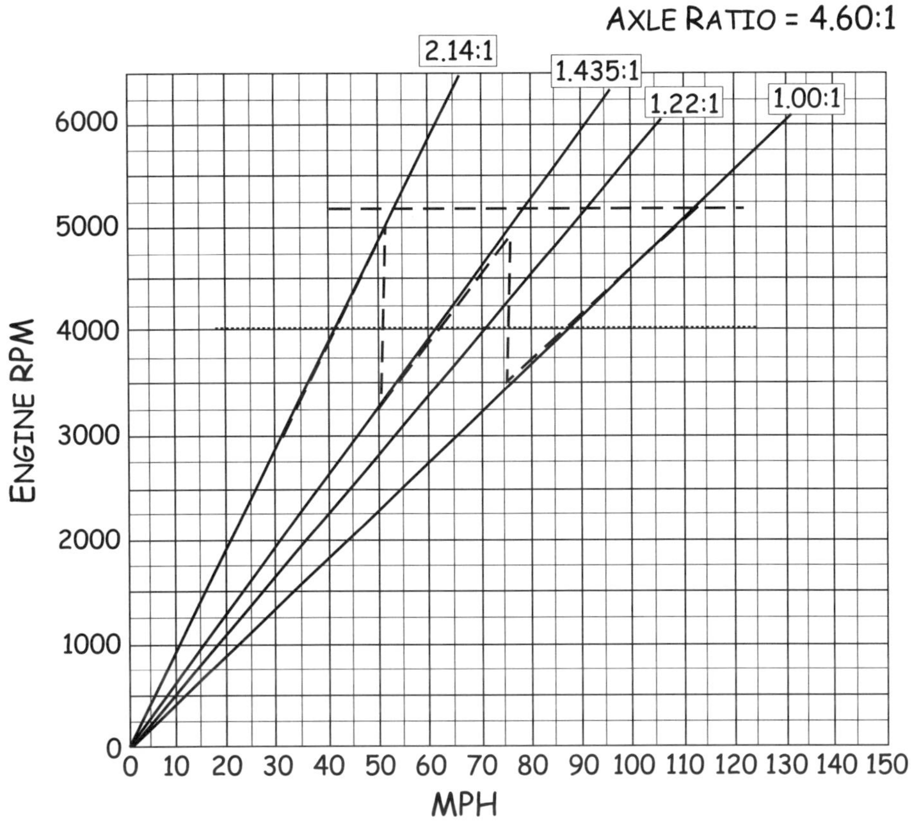

Illustration 11-3 Calculated speeds with 255 cu. in. engine.

April 16, 1934

Mr. Olen:

Mr. Brisko would like very much to test our racing car without a differential in rear axle to determine whether or not higher speeds can be attained when going into curves without the rear axle differential. In this connection the fastest rear drive cars do not have a differential, however, I understand all front wheel drive cars are equipped with differentials.

It is desired to make this test in order to determine what construction gives the greater speed and it is not our intention at this time to omit this differential it is merely for a test and our future actions, of course, will depend upon the results of test and in making such a test no one would know of the test, when it was conducted, nor would they know the results obtained. This information would be held strictly within our organization.

When going into a turn there is a tendency for the inside wheel to raise off the ground due to centrifugal force and it is therefore necessary to enter turns cautiously to prevent the inside wheel from spinning. This, however, should not apply to such an extend on a four-wheel-drive car as a car driven on one axle only and, therefore, this feature may not have any merit for our construction. It would, however, be interesting to make a test and know absolutely what results are obtained.

If it meets with your approval to make such a test I will have the necessary part made up and sent to Mr. Larson within the next few days.

Yours very truly,

H.B.Dodge
Chief Engineer
MIM
(hand written note): OK WAO

A statement of expenses from Dodge to Gause is full of derisory figures. The most stunning is $30 for an exhaust manifold formed and torched by hand out of sheet *stock.*

April 18, 1934

Mr. Gause:

The following is a list of expenses involved on our race car at Indianapolis. This work is being supervised by Mr. Brisko and Mr. Larson.

New exhaust manifold	$30.00
Pan under motor	12.50
Weld pan under body	5.00
Weld body and strengthen same	10.00
Cut louvers and put side panels and cut out panel for manifold	12.50
Lower water manifold	5.00
Prepare oil cooler	20.00
Total	$95.00

Attached hereto are a quantity of bills by Mr. Larson covering part of this work, also his expense account for trip to Indianapolis.

In order that this work may not be delayed will you kindly issue another check to Mr. Larson in the amount of $100.00. This should be sent promptly so that there will be no delay in work being done.

Yours very truly,

H.B. Dodge
Chief Engineer
MIM
Attach.

The original V-8 engine had been equipped with a 12-volt Bosch starter, as the Chancellor engine had been, and probably the Burden. Apparently it was sought to dispense with this convenience with the new Miller 255, a decision that was reconsidered after arrival at the Indianapolis Speedway.

COPY

April 20, 1934

Four Wheel Drive Auto Company
Clintonville
Wisconsin

Attention: Mr. H.B.Dodge, Chief Engineer

Gentlemen:

Confirming our wire sent you today and acknowledging your wire of the 20th. and your letter of the 18th, we recommend 12 volt starter type BNF 2.5, anticlockwise rotation, with 11 teeth, 8/10 pitch pinion, price $39.00 net. Our Engineering Department believes this type suitable for Miller 4 cylinder racing engine.

In our wire, we explain that it will be necessary for you to make your own holding sleeve, or order this from the Offenhauser Engineering Company, Los Angeles, California.

You will please note print 119, herewith enclosed, showing wiring diagram for this starter with battery carried in race car. Print 1321 gives wiring diagram for this starter when the battery is not carried in the car, but kept on the pit wall instead.

The distance between the pinion end and gear ring edge must not be less than 1/8 inch and not more than 5/32 inch for best performance.

The starting switch to be used with this motor is our part RS 82.

We can arrange to ship this starter by express not later than Wednesday, April 25, if your order is received promptly.

Yours very truly,
UNITED AMERICAN BOSCH CORPORATION
E.C. Peck (signed)

Edward C. Peck
Mfrs' Sales Division - Main Office
ECP:tld.
Encl. 2

The correspondence with Bosch ends here, but the starter resurfaces below.

COPY

April 21, 1934

Four Wheel Drive Auto Co.,
Clintonville, Wisconsin

Attn: Mr. H.B. Dodge, Chief Engineer

Gentlemen:

In response to wire received today, we enclose print giving mounting dimensions of starter type BNF 2.5/12 anti-clockwise rotation.

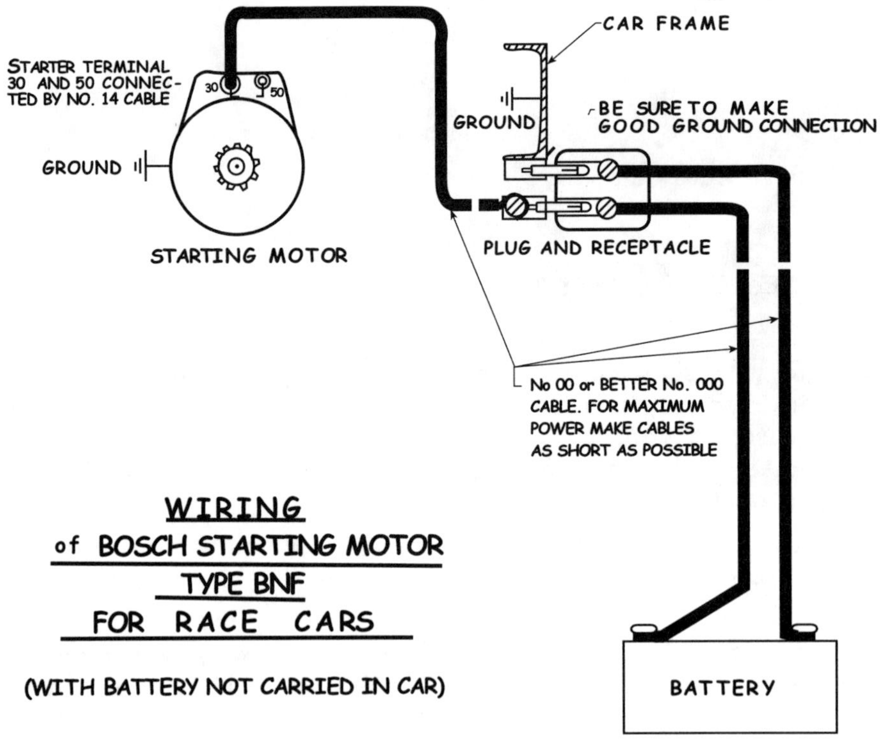

Illustration 11-4

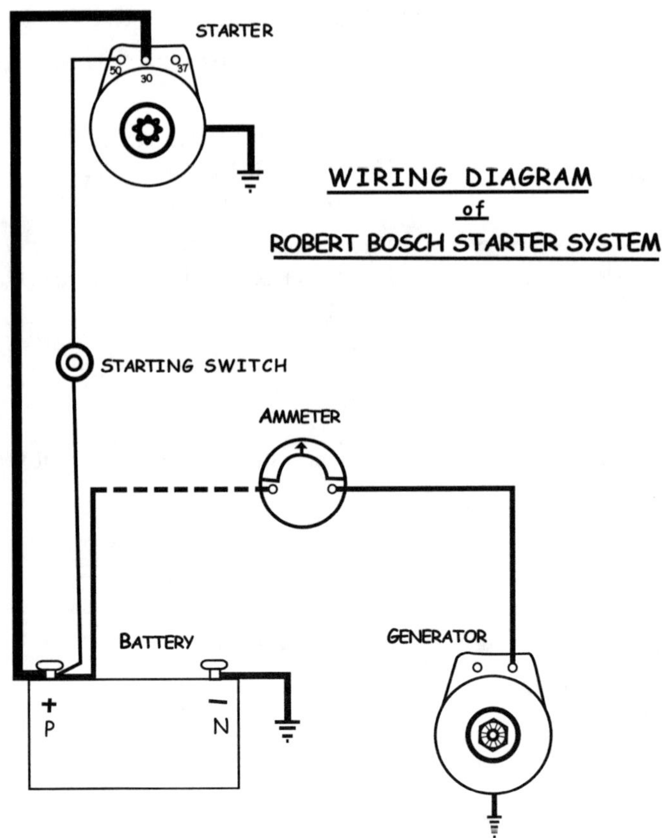

Illustration 11-5 Battery carried in car.

As mentioned in our letter to you of the 20th, this type starter motor is used in conjunction with a 12-volt battery. We will send you a torque curve early next week, when this has been made up by our Engineering Department.

For your information, please be informed that our experience with Miller racing cars indicates that this type starting motor will be entirely satisfactory for the purpose.

Upon receipt of this letter, please wire if we are to ship the starter, which as previously advised, can be forwarded not later than Wednesday, April 25.

Very truly yours
UNITED AMERICAN BOSCH CORPORATION
E. C. Peck (signed)

Edward C. Peck
Mfrs.' Sales Division
ECP:SM
Dictaphone
enc.

In 1934 the Company file included a completed copy of the Indianapolis Motor Speedway entry form for the 500. Brisko's riding mechanic was his brother Al.

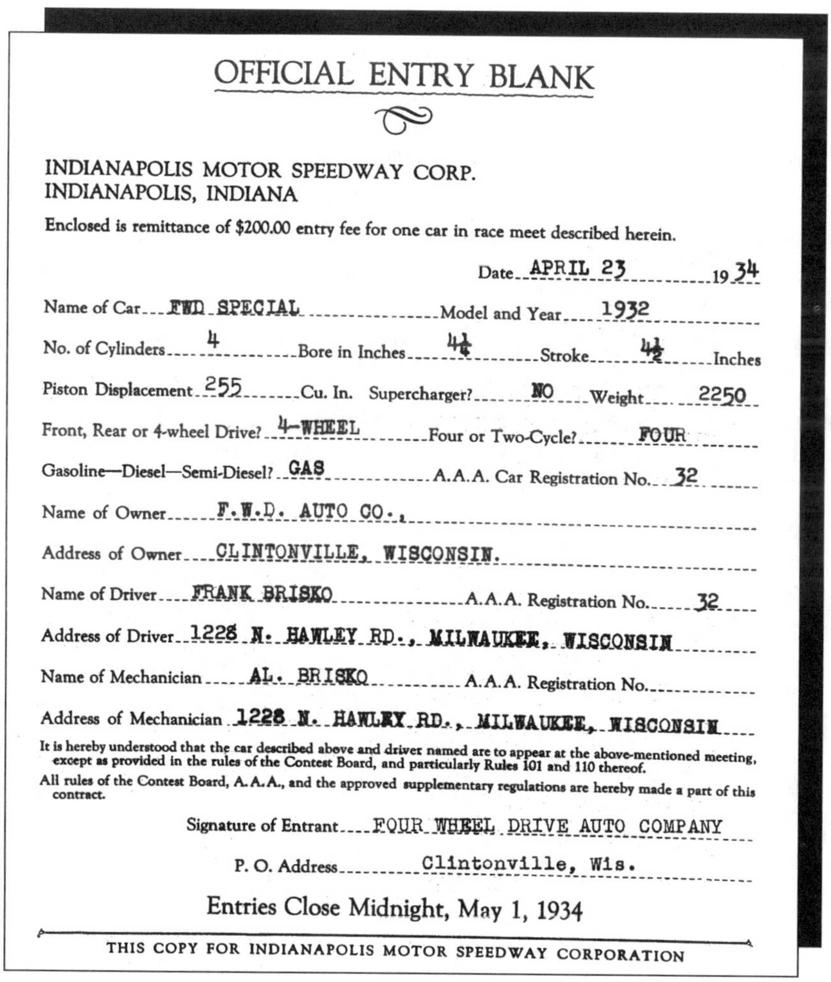

OFFICIAL ENTRY BLANK

INDIANAPOLIS MOTOR SPEEDWAY CORP.
INDIANAPOLIS, INDIANA

Enclosed is remittance of $200.00 entry fee for one car in race meet described herein.

Date APRIL 23 1934

Name of Car FWD SPECIAL Model and Year 1932

No. of Cylinders 4 Bore in Inches 4¼ Stroke 4½ Inches

Piston Displacement 255 Cu. In. Supercharger? NO Weight 2250

Front, Rear or 4-wheel Drive? 4-WHEEL Four or Two-Cycle? FOUR

Gasoline—Diesel—Semi-Diesel? GAS A.A.A. Car Registration No. 32

Name of Owner F.W.D. AUTO CO.,

Address of Owner CLINTONVILLE, WISCONSIN.

Name of Driver FRANK BRISKO A.A.A. Registration No. 32

Address of Driver 1228 N. HAWLEY RD., MILWAUKEE, WISCONSIN

Name of Mechanician AL. BRISKO A.A.A. Registration No.

Address of Mechanician 1228 N. HAWLEY RD., MILWAUKEE, WISCONSIN

It is hereby understood that the car described above and driver named are to appear at the above-mentioned meeting, except as provided in the rules of the Contest Board, and particularly Rules 101 and 110 thereof.

All rules of the Contest Board, A.A.A., and the approved supplementary regulations are hereby made a part of this contract.

Signature of Entrant FOUR WHEEL DRIVE AUTO COMPANY

P. O. Address Clintonville, Wis.

Entries Close Midnight, May 1, 1934

THIS COPY FOR INDIANAPOLIS MOTOR SPEEDWAY CORPORATION

Illustration 11-6

As in the past, the Company man assigned to the car was Herman Larson, who reported to Dodge. For those who find his handwriting difficult to read, on April 27 he wrote:

4/27 - 34

Dear Mr. Dodge

We only worked ten hours today on account of the cold weather, we only have a blow torch and two electric light bulbs for heating the garage. The temperature this morning was 36°.

We found valve guides broken in the same part where cup was broken, this guide has been replaced.

Cylinder block has been cleaned of all grit and valves reground.

In fitting the new valve cups we found them perfect in thickness at top the others are only off .001 to .003 not bad. [The caps undoubtedly were furnished by Offenhauser. Their fabrication is a very specialized art]. The whole car should be in perfect condition when we get it assembled this time.

If the weather is a little warmer we will be working late in order to make some tests Sunday, although it is not good to tow the car for starting.

If the starting motor come in Saturday, we will try and have it installed at once. you know these shops here also have N.R.A. [National Recovery Act; a New Deal measure to combat unemployment] hours, in mounting this starter all we have to have done bore out the old housing, have a sleeve made, and clip a band on rear end.

Got the price on 1934 tires today 7.00 x 18 tires $24.20 each tubes $4.75 each

The threads on the four wheel wing nuts are worn out will have to have a set before the race, but this will not delay our tests, will try and pick up a set here if the price is right.

There are no cars being tested at the present time on account of cold weather.

I do not believe the new valve cups will give us any trouble. Frank was a little worried for a while on account of the extra weight at top end what do you think of the cups I returned.

yours truly

H.V. Larson

H. V. Larson

Dodge responded to Larson on April 30th:

April 30, 1934

Mr. H. V. Larson,
Sheffield Inn,
Indianapolis, Indiana

Dear Herman:-

We have your wire stating that the Merz Engineering Company is proceeding with the mounting of starter. You did not, however, state that the starter had arrived. We have wired Bosch to find out shipping date but as yet have no reply.

You[r] wire stated that towing the car for starting has a tendency to loosen the splines in hubs. I cannot see how it would be possible to loosen the splines by starting the engine in this manner and if splines are loose possibly the material is soft. Would suggest that this matter be checked over at once and if any parts are required to be made over we will have to work pretty fast as time is now getting short.

We are most anxious to have the trial of this car run as promptly as possible for the reason above mentioned. If anything serious develops we will not now have time to make new parts for correcting difficulty. It is, therefore, most important that the car be tested thoroughly just as promptly as possible.

I mentioned in previous letter regarding establishing a credit for tires and if you have not already done so kindly have the Firestone representative wire in to the Firestone factory or possibly to Mr. Ziegler of their Milwaukee branch and obtain permission to sell tires and charge same to our factory. I am inclosing herewith a letter to their representative in Indianapolis which may be of service to you.

Yours very truly,
FOUR WHEEL DRIVE AUTO CO.,

B.B.Dodge
Chief Engineer

MIM
Inc.,

On May 2 Larson reported to Dodge:

5/2 -34
Dear Mr Dodge

The Merz Engineering Co worked until 12:30 this morning on starter mounting. we took off four hours for rest before going to the Speedway this morning. we had some fitting to do in order to mount starter but we have a good job

Mr Volmer from Bosch [whose main function at the Speedway was looking after magnetos] came in late today he checked the mounting and said it was O.K.

We tried to start with 12-Volt battery, but had to use18 volts in order to spin motor. [One wonders what the compression ratio must have been].

Am inclosing two sheets on weights of Racing Car--the total weight in each case checked close with front and rear also side--in weighing single wheels I had to correct them slightly

After taking our weights we warmed up our motor at our garage and blew out a couple of hose connections on oil line, we put in new hose and got out on the track at 5.30 o'olock and made nine laps--6 laps at 100 miles and three laps at 113 M.P.H. motor seemed to be in perfect condition we will check this in the morning

yours truly

H.V. Larson
H.V. Larson

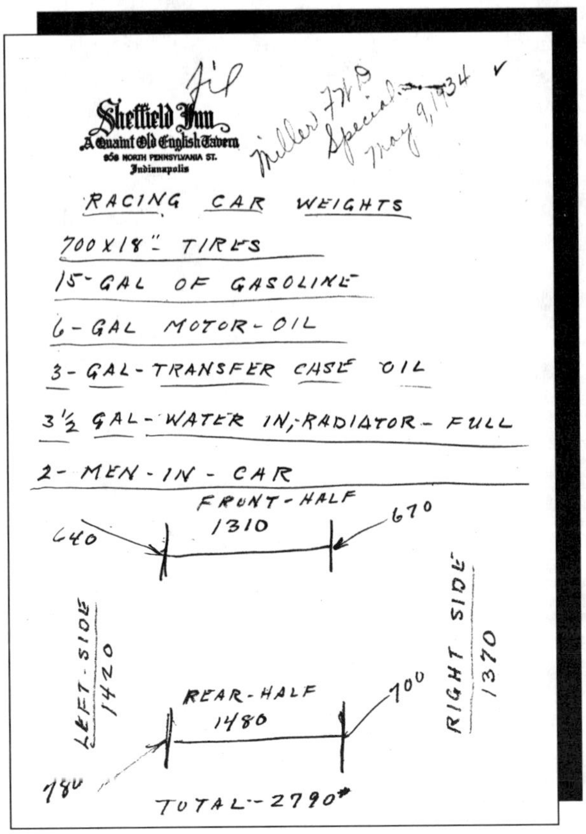

File
Sheffield Inn
A Quaint Old English Tavern
956 North Pennsylvania St.
Indianapolis

Miller FWD Special May 9, 1934 ✓

RACING CAR WEIGHTS
700 X 18" TIRES
15-GAL OF GASOLINE
6-GAL MOTOR-OIL
3-GAL-TRANSFER CASE OIL
3½ GAL-WATER IN RADIATOR-FULL
2-MEN-IN-CAR

FRONT-HALF 1310
640 | 670
LEFT-SIDE 1420
RIGHT SIDE 1370
REAR-HALF 1480
780 | 700
TOTAL--2790#

Illustration 11-7

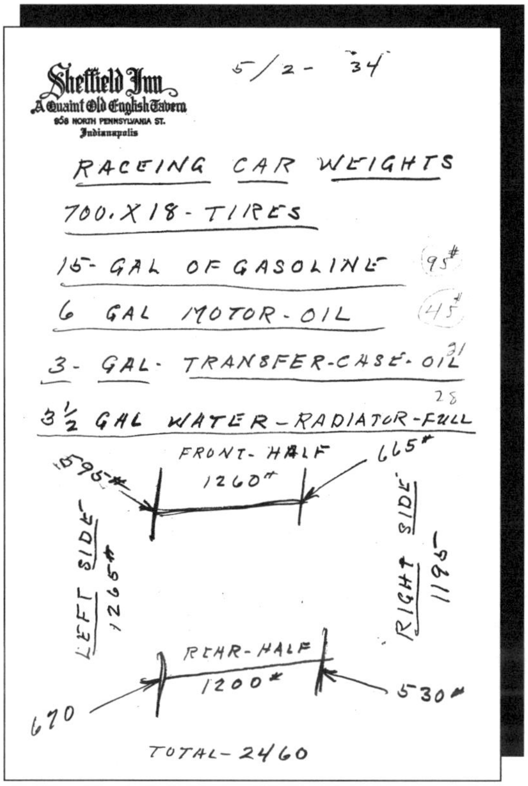

Sheffield Inn
A Quaint Old English Tavern
958 North Pennsylvania St.
Indianapolis

5/2-34

RACEING CAR WEIGHTS

700.X18-TIRES

15-GAL OF GASOLINE 95#

6 GAL MOTOR-OIL 45#

3-GAL-TRANSFER-CASE-OIL 31

3½ GAL WATER-RADIATOR-FULL 28

FRONT-HALF 1260# 595# 665#

LEFT SIDE 1265#

RIGHT SIDE 1195

REAR-HALF 1200# 670 530#

TOTAL-2460

Illustration 11-8

It is interesting to note that the ambient temperature rose from 36° to 79° F in the space of six days.

Also on May 2nd, Olen drafted a press release about the car, for distribution within the Company and to the press. He was careful to attribute the change of engines to a change in rules at Indianapolis Motor Speedway.

5/2/34

RACING CAR

New regulations adoped by the Indianapolis Motor Speedway Corporation have made it necessary to make material changes in the racing car. The entire engine lubricating system including reserve supply must not exceed six gallons and two quarts including expansion space and must be completely drainable. No further supply may be taken on during the race. No engine oil supplies will be allowed in the pits.

Capacity of fuel tank to top of filler necks including fuel lines and carburetor bowl shall be fifteen gallons. However, a tolerance of not more than two quarts will be allowed.

With the eight (8) cylinder motor it would not be possible to run the 500 mile race with six gallons and two quarts of oil and fifteen gallons of gasoline. It is necessary to average better than eleven miles per gallon.

The company was interested in a proper motor for the car and finally decided on the Miller four cylinder motor. Mr. Brisko was sent to LosAngeles and was successful in purchasing complete drawings and specifications of the motor together with a complete set of castings and forgings and out of it the company has built their own four cylinder motor. This reduced the weight of the car between 200 and 300 lbs.

The car is now at Indianapolis in charge of Frank Brisko and H. V. Larson where runs are being made each day and further refinements are being made in the car. The present indications are that both the oil and gasoline will be sufficient to run the race with this motor. The company was interested in this particular motor not only for the race car, but with the plan of having it for other use which will be determined as soon as the efficiency and reliability of the motor have been determined, either in the preliminaries or the final race. It was entirely built in our own plant.

In view of the unsatisfactory showing on two previous occasions, we desire to be extremely cautious and careful in our predictions this year. The foregoing is simply to let you know the present status of the racing car.

Walter A. Olen

Walter A. Olen

On May 7th Larson wrote:

5/7 - 34

Dear Mr Dodge

Today I got the body and under pan and radiator shell repaired, so all parts that are to be painted are now at the paint shop.

In repairing body Mr Ringling [Rigling] cut the body at both side and made a joint as per sketch, this in order to reduce the leverage at this point. I believe this will eliminate our Trouble he did a good job in welding all the cracks—also in patching and welding middle pan, at a cost of $15.00 I owed him $3.00 on the $75.00 contract. I paid him in full today

The Acme Radiator Co repaired the radiator guard today all the vertical bars were resoldered, and tied together by soldering a heavy wire at the top out of sight and the bottom ends are also tied together with a light angle shape and soldered The radiator has not been repaired we will get this Wednesday.

On the last test the handle from the oil control valve bounced off on the track some place, so will have one made up and welded on the valve stem

The top water manifold is cracked in two places on front leg, caused no doubt by rough track and three point engine mounting, we will have to make up a new one from heavier material, and cut all the legs and use hose connections between this should take care of all vibration, Mr Ringling will make up a new one for $12.50 the one we got cost about $60.00

Bill Cummings was out today with the Boyle 8 cylinder trying out some new carburetors he only made a few laps, and cracked the body, and had to pull out. The new carburetors were not so hot either

Wilbur Shaw run four laps and pulled in, for some leaks in water lines

Chet Miller was out with his 8 cylinder Studebaker, to test gasoline mileage, he run three warmup laps, then took all his equipment back to the garage

yours truly

H. V. Larson

H. V. Larson

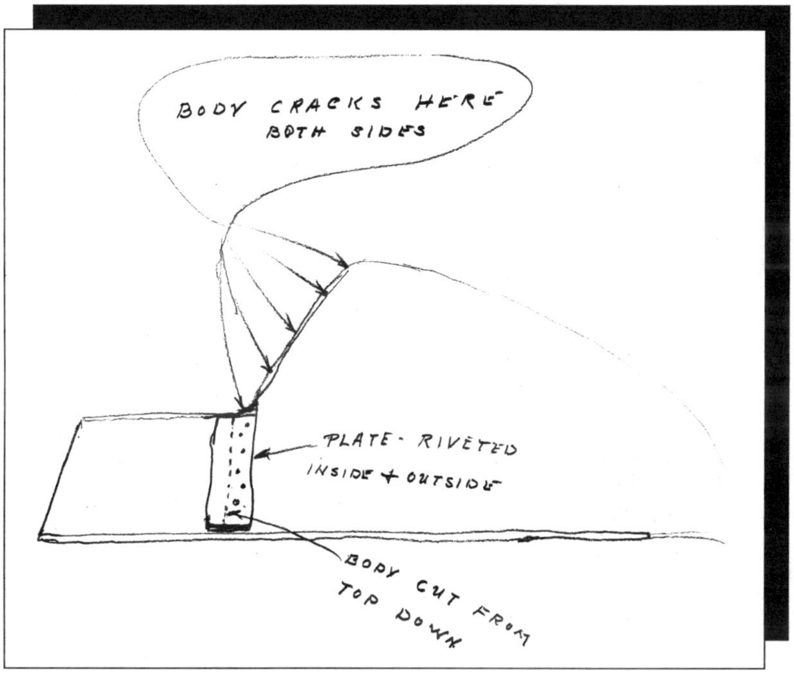

Illustration 11-9
Sketch of body repair.

On the following day, May 8th, the now-familiar lease agreement between the Four Wheel Drive Auto Company and Frank Brisko was executed. It was good only for the period May 16th, 1934, through June 4th, 1934.

THIS INDENTURE made and entered into this 8th. day of May, 1934, between the Four Wheel Drive Auto Company of Clintonville, Wisconsin, hereinafter called "Lessor" and Frank F. Brisko of 1228 North Hawley Road, Milwaukee, Wisconsin, hereinafter called "Lessee".

WITNESSETH:

That the Lessor hereby leases to Lessee and Lessee hereby agrees to rent from the Lessor, one racing car described and known as the "Four Wheel Drive Special" owned by the Four Wheel Drive Auto Company, for the term beginning May 16th, 1934 and ending on June 4,1934,at 5 o'clock P.M.

It is expressly agreed and understood that said racing car may be driven by lessee in the 500 mile automobile race to be run at Indianapolis Speedway on May 30, 1934.

The lessor is to furnish all repair parts, lubricating oil, grease, gasoline, and tires required during the term of this lease, to put and keep said car in condition and to participate in said lease.

IN CONSIDERATION THEREOF, the lessee binds himself to pay as rent for same a sum equal to 50% of the gross prize winnings or earnings of said car earned as a result of the lessee participating in said race, said sum to be due and payable to lessor at the office of the Indianapolis Motor Speedway Corporation immediately upon payment of said prize money or earnings.

The lessee further agrees that immediately after the 500 mile automobile race on May 30, 1934 at the Indianapolis Speedway, he will deliver said Four wheel Drive Special car to the Four Wheel Drive Auto Company at Clintonville, Wisconsin, said delivery to be made at the point designated not later than 5 o'clock P.M. June 4, 1934, unless otherwise agreed upon in writing between lessee and lessor.

It is further agreed that lessor shall not be liable to said lessee for damages of any kind or from any cause whatever to the persons or property of lessee, or any person claiming by, through or under him, and lessee agrees to defend at his own expense all damage suits which may be brought upon claims relating to said cause or the use thereof by the lessee.

IN WITNESS WHEREOF, the parties hereto have hereunto set their hands and seals the day and year first above written.

	FOUR WHEEL DRIVE AUTO CO.
Attest:	Lessor
(signed) Helen Neller, ass't Secy.	(signed) Frank Gause, secretary
Witness:	(signed) Frank Brisko
(signed) H.V. Larson	Lessee

Brisko wrote to Dodge on May 13th, saying:

May 13 - 1934

Dear Dodge.

Received your letter of the 11th and expect the cups will be here in the morning I have the cylinder block off as I had to give the pistons more clearce. So will keep it off till I get the new cups in.

If the cups are in by 9 a.m. we will do some running tomorrow as I want to run a last 25 miles so I know how far the gas will go to qualify.

We won't change the gear ratio as this would ... to much for the race because after You qualify You must keep the same gear in.

So I will leave it alone so we can get better mileage this way.

Herman is writing You in his report what Meyers done with his cup.

Yours truly Frank Brisko

And Herman's letter said:

5/13 - 34

Dear Mr Dodge

After talking with Louis Meyers [Meyer], today and look at cups in his car, we decided to wait for the 2512 [steel alloy] valve cups. we believe we can save time this way, and not take any chances, with old cups unless we have to.

Meyers has the Keatons [?] steel cups, with a hard surface welded on top he says the Brinell hardness of top is now 900. he will have an extra supply of this welding rod in a few days in case we need some, but I still believe our 2512 steel will do the trick.

The welded cups in Meyers car look very good no scuffing. but they have only been run 75 miles below 100 miles per hour

If we get the 2512 cups in the morning we should be able to make at least one test Monday

We got our top water manifold today, radiator and shell installed, the body we will pick up Monday morning

yours truly

H.V. Larson

H. V. Larson

On a date that we may never know, Frank Brisko was engaged by the Four Wheel Drive Auto Company in the capacity of race car engineer. As we shall see, he was sent to Los Angeles, where he obtained the drawings and most of the parts for building a Miller 4C255 engine, which he did at the Clintonville factory. With Larson, he took car and engine to the Indianapolis Speedway, becoming the car's lessee on May 8th. On May 14th he resigned from his apparent position on the Company payroll, freeing the Company of all liability for what might go wrong during the adventure of all concerned in the 1934 Indianapolis 500-mile race.

May 14, 1934

Four Wheel Drive Sales Co.
Clintonville, Wisconsin

Atten: H.B. Dodge

Gentlemen:

I hereby resign my position as race car engineer for the Four Wheel Drive Auto Company to take effect on May 15th, 1934.

Very truly yours,

Frank Brisko

Frank Brisko

Then came the big race, which we will deal with in a moment. The Company file labeled "Race Car 1934" contains only two other documents. The first, from famous mechanic Cotton Henning to the company's Mr. Higgins concerns the type of front outer U-joint used on Wild Bill Cummings's winning Miller 4C220 FWD. It really gives food for thought:

June 17, 1934

The Four Wheel Drive Co.
Clintonville Wis.

Dear Mr. Higgins

In regards to your letter of June 13th about the type universal joints Cummings used in the front drive, were Rzeppa joints, made by the Lear Grinding Machine Co. Detroit Mich. I had them made special a year ago. A very nice joint, and the only one I have found that is really safe for front wheel drive. Hoping this is the desired information. I remain

Very truly yours,

Cotton Henning.

The last document of the year is a scientific one and is as fascinating as it is precious. In the fourth paragraph, "46_" should read "46%." The objective testimony of Brisko's performance for the first 250 miles is very nice to have.

(COPY)

Firestone Tire & Rubber Company
Milwaukee, Wisconsin
August 4, 1934

Mr. G. H. Billings, Purchasing Agent
Four Wheel Drive Auto Co.
Clintonville, Wisc.

Dear Mr. Billings:-

You will recall sometime ago that you requested information from me as to the amount of tread wear on your FWD racing car at the recent Indianapolis races - I took this matter up with Mr. Brownie Carslake and have received his reply which was delayed to some extent owing to the fact that this data was not completed until recently and I am showing below complete data covering the eight fastest qualifying cars at the races:

The qualifying trials for the 1934 Indianapolis race showed the following tread wear loss in percentage -- (right driving wheel):

DRIVER	QUAL.SPEED	RELATIVE AMOUNT OF TREAD WEAR (%)
Petillo	119.329 MPH	154%
Shaw	117.647	131%
Brisko (4 wheel)	116.894	54% RR
Cummings (front)	116.116	89%
Gardner	114.786	126%
Barringer	113.859	83%
Shafer	113.816	69%
Miller	113.307	95%
Average	115.719	100%

During qualifications the track was dry and abrasive and tread wear was abnormally fast.

The above shows that the treat wear loss on the four wheel drive car was 46_ better than the average of the eight cars, inspite of the fact that Mr. Brisko was more than a mile an hour faster than the average speed of these cars.

These figures are the most accurate for comparison purposes because speed and track conditions are the major factors in determining tread wear, and while Mr. Brisko's average speed for the actual race was low, his speed for the first 250 miles while the track was dry and abrasive was 104 MPH during most of which time he led the race, so his tread wear for the race, while not excessive, was greater than his average speed would indicate.

However, I wish to call your attention to the fact that all of this is very confidential data and should be used only for engineering purposes and should not be advertised or published in any way, and I believe you fully appreciate the importance of our co-operating with you to this extent and will keep this matter confidential.

Very truly yours,
FIRESTONE TIRE & RUBBER COMPANY
(Signed) W.A.Ziegler
MFRS. SALES DEPARTMENT

WAZ:T

Chapter 12

Indy '34—
A Fine Technician and His Finest Ride

Far from being your run-of-the-mill throttle jockey, Frank Brisko was a gifted and creative mechanic, machinist, and practical engineer. Much of what he drove was built with his own hands. He served his apprenticeship on the I.M.C.A.* dirt, before graduating to the A.A.A. league and Indianapolis. His rookie ride there was in a front-wheel-drive Miller of which he seemed to be the owner. He qualified twenty-ninth but finished eleventh, almost in the money. He skipped the next year, but, as was noted early in Chapter 10, in 1931 he began his long liaison with the original outboard-brake front-wheel drive Miller. What that particular mechanical monument meant to him we shall never know, but he and it constituted a pair. He remained its owner for a decade, campaigning it loyally, although it did not become even slightly competitive until 1940 when, massively transformed, the relic finished the Indianapolis 500 miles in ninth place. He rarely had an opportunity to show what a good driver he was. He seemed to get his satisfaction out of being part of the group experience; out of going along for the ride. But if and when he got within striking distance of a really good position, he could hang in there with the best.

As we have seen, in 1934 and probably for a good part of 1933, Brisko served as a key member of what amounted to the Four Wheel Drive Auto Company's racing department. Company people, who avowedly knew nothing about the racing field, depended on this home-state young man for expert guidance in many domains, from product conception and design, to artisan machine-shop practice, to campaigning and driving the race car. He had a particular gift for engines. After having lived with Miller power packages for a few years, he began to have his own ideas on how they should be improved, becoming a builder of custom racing engines in his own right. Perhaps it was with the 4C255 that he built on the Four Wheel Drive Auto Company's premises that this activity of his became a continuing specialty. He built mostly for the dirt track and the sprint-car field, but Indianapolis entry lists for 1936, 1937, 1941, and 1946 show one Brisko engine in each event, while there were two in 1938, 1939, and 1940. When Buck Boudeman brought the Miller front-wheel-drive Special home in around 1984, he telephoned Brisko to learn what he might remember about the machine.

---

* International Motor Contest Association.

Illustration 12-1 A harbinger of the coming Offy era was this early 4C255. It was built up by Brisko in Milwaukee and Clintonville to replace the underdeveloped V-8 in the FWD Special. Brisko, a carburetion specialist, shunned downdrafts. (RMRC)

Illustration 12-2 This Floyd Dreyer-built dirt-track one-man car is of about 1934 vintage and forms part of the David Uihlein Collection. Its engine is not an Offy, but a Brisko. Whereas Fred Offenhauser remained true to the barrel crankcase, Brisko, like Miller, had decided in favor of the split case with conventional main bearings. This is a 4C268. Brisko obtained certain parts, such as gear towers, from Offenhauser, made the rest himself. (David Uihlein)

"He remembered everything," Buck told me ten years later. "He had it all on the tip of his tongue. He knew what the valve lash was, the valve-guide clearance, and all that. He built the engine, and claimed that it was the first 255. When Miller went broke, Brisko went to the Coast and rented the patterns and drawings from Loynes. He told me that he had the complete patterns, which were not yet finished. So he knew that nobody else had gotten a casting off them because he had to finish them. He built an engine for himself, then I think one for Louis Meyer. He built those at the Four Wheel Drive Auto Company plant.* I think that Fred Offenhauser made the crankshafts and maybe the oil pumps, and Ed Winfield ground the cams. I think that Brisko may have gotten the gear towers from Offenhauser as well. Frank machined the crankcase, cam housings, cylinder blocks, and so on, and put them all together."

As we have already learned from Company correspondence, Offenhauser obtained his 255 drawings from Brisko (p. 200, bottom), who also had physical possession of the patterns, which he said he had to "finish." How, then, were the 255s that powered the Lou Moore/ Scully and the Lester Spangler/Harry Hartz cars and others at Indianapolis in 1933 built? This is simple enough: They were built well before the U.S. marshal padlocked the Miller plant's doors. Why did Brisko believe that the patterns were unfinished? Well, perhaps because between the closing of the plant, the sale, and their eventual delivery to Brisko the patterns had been so damaged as to seem to be unfinished. Why were the block castings not made at the McCauley Foundry on the West Coast, as always had been done before? McCauley was one of the chief creditors who had lowered the boom on Miller; the situation may have had its delicate aspects. And, the Four Wheel Drive Auto Company probably had its own foundry, and its nearby engine suppliers certainly had theirs. Bunny Phillips was very precise in stating that the Scully 255 was number five of that series. How could Brisko think that his 255 was the first? He was not a man to deal lightly with facts. This may be explained by something that we do not as yet know, such as:

When I first interviewed Offenhauser in August 1953 he told me, in the context of having championed the four-banger against Miller's will, "I built a 255 for a local client while Harry and Leo were in Detroit on the Gar Wood job." Which is to say in August and September of 1931.

Under these conditions of unauthorized initiative, that engine is unlikely to have been engineered from scratch, but merely to have been a 220 stroked and bored almost 1/4 in. to the water. It happened to go and, in spite of Harry Miller's disapproval, it attracted a few other buyers. This was an unsuspected market segment that was worth cultivating, and the decision was made to do so in a serious manner. "Go ahead, Leo; draw it up the way it should be." Which he did, after which work on the patterns was begun. Due to the events that we know, they arrived in Loynes' hands unfinished. In this way, or something like it, Brisko came to build the first 255 to be engineered as such, and not merely a stretched 220. Readers interested in the roots of the Offy will do well to consult *The Miller Dynasty*, by Mark Dees, page 390, and especially the fourth paragraph.

---

* According to Dave Uihlein, the major machine work was done by Brisko in the Marchese brothers' shop in Milwaukee.

I attempted to contact Brisko in 1983. He passed my letter to Dave Uihlein:

Mirail, 19 Aug. 1983

Dear Mr. Brisko:

I have recently been given your address by Al Bloemker, of the Speedway.

In a story on Miller front-wheel drive which I have written for AUTOMOBILE QUARTERLY I have devoted a lot of space to the history of your old car--the one which Benny Hill refused to drive in the 1925 500.

I would like to know if you know what finally became of that great old machine and if you possibly know if it survives today.

Thank you very much for giving your attention to this request, and best regards.

Sincerely,

European Editor, AQ
GRIFFITH BORGESON'
MIRAIL
8240 LA MOTTE D'AIGUES
FRANCE

September 1, 1983

Mr. Griffith Borgeson
Campagne Mirail
La Motte d'Aigues
84240 La Tour D'Aigues
FRANCE

Dear Mr. Borgeson:

Frank Brisko has passed on to me your letter of August 19th inquiring about the 122 Front Drive "outboard brake" car.

I thought I had told you that I found the car in Chicago and bought it about two years ago. We removed Frank's body (called the Boxar Tool Special) as well as his six cylinder engine. I am in the process of restoring it, and have made new 20" wheels, completely rebuilt the body to its original configuration and acquired Carl Marchese' 122 Miller engine so that we have the makings of restoring a totally original 122 Front Drive.

Hope this answers your questions.

Sincerely,
(signed) DAVE
David V. Uihlein
bar

In 1994, I asked Uihlein for information on Brisko and was rewarded with this trove. It included the present location of the Four Wheel Drive Auto Company car's original V-8 engine.

April 27, 1994

Attention: Griff Borgeson

Dear Griff,

Not only did I know Frank Brisko, but he came to my museum and we had several very nice conversations. I am sorry to say I didn't tape any of them.

He was a great rival of my good friends, the Marchese brothers, and only passed away a year or two ago at 89. In his younger days, he drove both on the dirt and Indy. Somewhere along the way, after 1929, he designed and built big 4 cylinder Miller type engines and finally a 6 cylinder which, incidentally, I just traded to Mark Dees. Both the 4 and the 6 utilized the 151 Miller marine gear tower which Carl Marchese always referred to as "Miller marine dimensions." Obviously, he built a bigger block, a bigger, longer crank case, cams and covers. Incidentally, I have a Brisko 4 cylinder engine of 268 cu. in. in one of Floyd Dreyer's very original sprint cars (1/2 and 1 mile dirt car). Frank looked at this engine and without hesitation claimed that it was Wild Bill Cummings engine of 1934. Brisko's design didn't utilize a barrel-type crankcase, but the traditional split case--quite frankly, it appears to be more durable. Enclosed is a picture of this car, and I will take the hood off and get you some engine pictures.

With regard to the 4WD, which he drove, I believe in 1934, my friend and his rival, Carl Marchese, who was always very complimentary of his rivals, said (speaking of Indy) "Frank Brisko led the race and I don't mean for a couple of miles, but he led it for 300 miles!"

I am sure you know that the original engine was a V8 Miller of, I believe, 308 cu. in., and that this engine exists in the collection of Bill Goodwin, the undertaker at Frankfurt, Indiana. You and Bob Rubin and Goodwin should know that I have the original 4 Miller carburetors from this engine. When Goodwin bought the engine, they had been removed, and came to me with all the rest of the Marchese Miller parts, and I am told that Bunny Phillips also has an identical V8 Miller.

A number of years ago, Frank moved up north to a little town called Ojibwa, where he spent the rest of his life. He devoted many years to gas savers and later to anti-pollution devices, and, while I cannot say whether they were successful, I do not believe any of his inventions were accepted by the large automobile companies.

I also enclose a picture of the present status of the Outboard Brake car, and you will note that it is far along, approaching completion. Danaher, the engine man, ran the engine last week and it turned 5100 rpm's. He said, "the most exciting sound of any engine we have tested. It is very vibration-free." This 122 is the last great effort of the Marchese brothers, whereas they changed the firing order, eliminating the cause of broken crankshafts. I hope to have the car finished and back in America by the end of July.

All good wishes to you, with kind regards from M-M and me.

(signed) DAVE

cc: Bob Rubin

P.S. As I recall more on Brisko, I will write you.

P.S.S. You should have some pictures of the Boxar Tool Special, which was the Outboard Brake chassis (brakes moved inboard) in its final form, with 4.5 litre, 6 cyclinder Brisko engine. I also sold the body to Mark Dees. I suspect that he might wish to recreate the car, as he now owns the engine and the body, whereas my great interest is, of course, in its "original" form.

P.S.S.S. I believe that the Olin family owned the 4WD Co. when the car was built, approximately 1931 or 1932. I worked for Bob Olin in 1948 when Wisconsin had its Centennial, but I do not know whether any members of his family are still around. The company is currently owned by some conglomerate.

(The pictures mentioned above will be mailed to you.)

Prominent Miller collector Charles F. Davis also knows the Goodwin-owned V-8 well and told me in April 1995 that it looks great and is complete except for the Uihlein carburetors. He sent a couple of snapshots of the engine and wrote:

"It was taken out of the car by Frank Brisko at the Marcheses' shop in Milwaukee in the winter of 1933 (?). The engine stayed at the shop until Goodwin bought it many years ago ...

Illustration 12-3 The ex-Four Wheel Drive Company V-8 engine for many years has formed part of the collection of the Goodwin Museum in Frankfort, Indiana. Inlets are properly blanked off; magnetos are absent. (Charles F. Davis)

Illustration 12-4 The ex-Miller V-8 at the Goodwin Museum, showing fly-wheel, ring gear, clutch, crankcase breathers, bell housing designed to mount Bosch electric starter. (Charles F. Davis)

"At one time I had the transmission that was on Bunny's V-8. It was complete with bell housing, flywheel, clutch, pedals, U-joint, and coupling. I traded this to Jim Brucker ... about ten years ago. He wanted the transmission for the Big 8 car he had. He has just sold the car with the 4WD transmission to Richard Freshman.

"Frank Brisko was a great sportsman. He was a good driver, mechanic, and mechanical engineer. I personally knew Frank and he helped me acquire a lot of the Miller stuff that I now own. I could not have done it without his help. He designed, built, and raced a 6 cyl. engine patterned after Miller 4 cyl. 151. Mark Dees has one of his engines in his collection. Frank also designed and raced a 91 c.i. 4 cyl. engine with hydraulic valves (no cam). He was quite successful with this engine in the '40s."

It is to be noted that, after the 1933 Indianapolis race, Mike Boyle had a Brisko-built, Miller-derived 4C220 installed in his Miller front-wheel drive Boyle Products Special. Wild Bill Cummings won the1934 Indianapolis 500 with this source of power and in 1935 qualified fifth and finished third with it. Brisko made other engines of this type, all with split crankcases, for which he had to have his own patterns. In doing so, he followed Miller practice, which began with the V8/308 of 1932.

The Unchaining of Man and Machine

This was the year of the four-banger explosion. Of the thirty-three qualifiers at Indianapolis, sixteen or forty-eight percent were of this configuration. Four were Miller 255s, Kelly Petillo's being the fastest of all at 119.329 mph. In the middle of the front row was Wilbur Shaw, with an 8C249 Miller, at 117.647 mph. On the outside, at 116.894 mph, was Frank Brisko in the four-wheel drive, with its Clintonville-built Miller 255. Brisko's engine was one of the three in the starting field to run updraft carburetors; the reader may speculate as to why. And Frank was one of the six who went the whole 500 miles without a tire change. More than that, he was the most economical consumer of rubber in the race, as Firestone's Ziegler testified.

Brisko was brilliant, moving into second place at the end of the first lap. Quickly thereafter he took over the lead, which he held for almost 200 miles! His first stop for fuel permitted Mauri Rose to take over the leading position. Back in the fray, Frank hacked his way past car after car, regaining the lead at about 275 miles. Then another pit stop, after which he got no higher than fourth, finishing ninth. He demonstrated what he was capable of, given a going machine. In his championing of the four-banger he had been perfectly in tune with the times. Fours took the first four places at Indianapolis, and six of the top ten. Newly made Miller Fours would receive nameplates that read "Offenhauser Engineering," and so the Offy era was born.

Illustration 12-5 What the company called "Race Car 1934" gave Brisko the greatest ride of his career. That year's Indianapolis 500 also brought the second-finest performance ever by a four-wheel-drive Miller. (RMRC)

Chapter 13

Tripoli and Berlin '34: The Swan Song of the 4WD Mule

What had begun as Scully and Phillip's Bugatti Service, with racing as a sideline, became increasingly involved in the latter. The emphasis on the ultra-scarce Molsheim marque faded, but the partnership remained intact. Enjoying the benefits of De Paolo's rich experience and young Lou Moore's driving ability and mechanical astuteness, they found themselves at the core of a racing organization with a great deal of promise. Frank and Bunny still yearned to show the Stars and Stripes in Europe, and Frank's well-to-do brother, Al, was lending his support to their enterprise. De Paolo had raced in Europe and had loved it. First, he was riding mechanic with his uncle, Ralph De Palma, on Ballots. Then, following his record-shattering Indianapolis win in 1925, he was invited to join the Alfa Romeo team for the 497-mile Italian Grand Prix at Monza on September 6th. He spoke English with elegance and verve; his Italian was folkloric brocolino; Brooklynese. He had the time of his life at Monza, bringing his P2 Alfa home in fifth place, while his teammates, Brilli Peri and Campari, finished first and second. In his 1935 book, *Wall Smacker*, my old friend Pete De Paolo told how he and Lou Moore just happened to be invited, by His Excellency the Governor of the Italian colony of Libya, to participate in the glamorous Grand Prix of

Illustration 13-1 My first ride with Uncle Ralph DePalma. A boyish Pete De Paolo in a 3-liter Ballot at Beverly Hills Board Speedway, 1920 (GBC ex PDeP)

Tripoli on the 6th of May, 1934. Knowing Pete, it is natural to visualize him having worked long and imaginatively to bring this about. And of course it was the fruition of a long-time dream shared by Scully and Bunny.

The history of the 327-mile race on the very fast Mellaha Circuit is described in an excellent 1994 book, *Grand Prix Tripoli 1925–1940*, by Valerio Moretti, and is told in English as well as in Italian and French. It opens thus:

"Although there were 14 editions of the Tripoli Grand Prix, only the last eight earned the event international renown. The reason for this success, after a very stuttering start, undoubtedly lies with the [Italian] national lottery which was later attached to the race. The lottery was tremendously successful and turned the Grand Prix into the Corsa dei Milioni; The Race of the Millions (of Italian lire)."

Illustration 13-2 At Tripoli in 1934 Pete De Paolo treated governor of the colony, Marshal Italo Balbo, to a few laps of one of the world's fastest circuits in the 4WD mule. (GBC ex Ernie Olson)

During this period, the Italian colony in North Africa was under the sway of a charming ex-thug and hero of transatlantic aviation, young Marshal Italo Balbo. He used the Tripoli Grand Prix as a stage set for pageantry that made the race and the colony glitter in the world's eyes. As already noted, it rivaled Indianapolis as the richest race on earth.

De Paolo had been at his peak in 1925, when he and his Duesenberg were the first to win the Indianapolis 500-mile race at an average speed in excess of 100 mph. Their 101.127 mph was not exceeded until 1932, and that achievement commanded enduring and universal respect. Pete wore this distinction with good-natured modesty. The year after he won it, and still driving a Duesenberg, he finished fifth in the Indianapolis 500. In 1927, and driving a front-wheel drive Miller, he lasted only thirty-one laps at the Speedway, but he did win three A.A.A. National Championship races on other tracks. After that, he spun his wheels until his retirement in 1930, after a last fruitless effort at Indianapolis. So he was hardly at the top of his form when, on April 7, 1934, he, Lou Moore, and ace mechanic Ernie Olson sailed from New York for sunny Napoli and even sunnier Tripoli. De Paolo, dispensing his Italo-American argot, was lionized by all, starting with Marshal Balbo, who did two laps of the 8.14 mile circuit at his side in the four-wheel drive mule, then raced it in his personal airplane. De Paolo and Moore were Balbo's dinner guests in the Governor's Palace, along with fine European drivers Varzi, Trossi, and Chiron. The Governor turned out to be as informed about auto racing as he was avid, and the evening was an unforgettable one. Balbo, one of the few people in the colony who spoke a smattering of English, crowned the evening for the Americans by personally driving them back to their hotel. Pete did not specify what he drove, but it had to be an Alfa to suit the image of its charismatic owner.

Illustration 13-3 The first driver to average more than 100 mph for 500 miles was Pete De Paolo in this "sidewinder" supercharged Duesenberg (GBC ex PDeP)

There were no German entries in the big race, which really was an Alfa Romeo festival. There were four works/Scuderia Ferrari P3s and six privately entered 8C2600s. Three Bugattis, eight Maseratis, and the two Millers brought the total menu to a thundering twenty-three cars. Moretti reports that the four-wheel drive was able to lap at 162 mph, "a little faster than the estimated top speed of Piero Taruffi's 16-cylinder Maserati." Both the Millers were markedly slower, however, than the P3 Alfas, which De Paolo was the first to acknowledge. Still, he and Moore were determined to defend the American colors to the best of their ability. Speaking of which, Moretti records that Pete's pit was literally draped in Italo-American bunting. De Paolo finished sixth and Moore seventh. This was a remarkably honorable performance by cars with brakes, gearboxes, tires, and a number of other elements that had not been conceived for this sort of racing. They both were faster than Rene Dreyfus's Type 59 Bugatti. Not bad for a couple of innocents from abroad.

Beyond that, they were among the nine cars out of the original twenty-five that lasted the entire distance, an attrition rate of sixty-four percent. Bunny Phillips said that both cars went through the épreuve "without a hitch." Somebody had been doing a lot of right things, and very likely it was super-mechanic Ernie Olson, who had been around the four-wheel drives and their intimate problems since Indianapolis 1932. Who had fixed the mule's transfer case, and how? And how come the oil-overheating V-8 ran so reliably? We may never know. Scully's No. 5 255 in Moore's car, of course, just ran like a train.

The next Grand Prix of the season was Germany's Avus Grand Prix, held on a sort of super speedway consisting of two parallel, long straightaways joined by a slow curve at each end, and 5.95 miles to the lap. Pete and the four-wheel drive mule happened also to be booked for this event, to be held in the Berlin area on May 27, 1934. After further attentions lavished on the Americans by their charmed hosts in Libya, they shipped back to Naples, where Moore and Olson grabbed the luxurious S.S. Rex for New York. According to De Paolo, they arrived at the Indianapolis Speedway with their car just three days before the 500. Lou and Ernie must have torn down and re-sanitized the car on shipboard, since they achieved the impossible feat of tuning and qualifying the car, and then finishing third in the big race again, all in that ridiculous space of time. Scully's No. 5 255 had struck again.

Race day at the AVUS Grand Prix was marked by pouring rain. De Paolo said that while any speed was possible on the straights—he mentioned 170 mph—the turns at each end had to be taken at around 30 mph. Since his Miller's brakes were useless in the wet, he had to back off very early and downshift in order to stay in the race at all. It was for fifteen laps, 182.4 miles. De Paolo was up against not only his Alfa and Mazzeratti (his spelling) rivals from Tripoli, but also the Germans, on their home turf. As it was, Mercedes-Benz defaulted at the eleventh hour, but that still left the scary Type A Auto Unions. De Paolo wrote, "Running in third place with the race half over I heard a terrific smashing bang under the hood of my car. Two connecting rods had let go, practically cutting my engine in half. This forced my withdrawal from the race, to my disappointment, having had a wonderful chance of winning." The mortally wounded mule was shipped back to Los Angeles by slow freight. Racing driver and historian Taso Mathieson, who lived much of the racing history of the thirties at its source, made the following observations:

PARIS. 3rd August 1981

Dear Griff

Thank you for your letter of the 25th July which I found awaiting me when we returned to Paris after a week in the country. Thank you also for the De Palma "Basic Fact" sheet, To my mind there is no doubt whatever that, if Ralph compiled this sheet, he included every race that he had won from1 to 500 miles, and then possibly a few imaginary ones as well. A pity to spoil his splendid record by such a stupid and obvious exaggeration.

Now regarding Peter de Paolo. At both Tripoli and the Avus the 4WD Miller was nothing like as fast as the Alfa Romeos nor did it have such good brakes, fine car though it was. He claims that he finished 6th at Tripoli but the official records give him 7th place, while at the Avus he says that he was third when a couple of con rods broke in the Miller engine.

There were only eleven starters at the Avus. Three Auto-Unions, three Alfa Romeos (one of these driven by Moll, was streamlined), two three litre monoposto Maseratis, a 2.6 biposto "Monza" Alfa Romeo, an unidentified car, and Pete's Miller.

The report of the race that I have states that the Auto Unions and the Alfa Romeos fought out a battle ahead of the remainder of the field the Maseratis and the Miller being fairly matched for speed but the brakes of the former gave them a very decided advantage over the Miller. Stuck (Auto-Union) led easily for three quarters of the race before retiring, but his team mates, Momberger and Prince Zu Leiningen, were not upto his ability in handling such a difficult car as the Auto-Union and were harried by the Alfas Leiningen going out near the end letting Moll lead to the finish followed by Varzi (Alfa Romeo) Momberger (Auto Union) Howe (Maserati) and Nuvolari (Maserati). Poor Pete says he stayed with the leaders while he was running but I am sure that he was never higher than 6th at any time during the race and never within striking distance of the leaders. It is true that the track was wet at the start but the road dried out early on as can be seen from the photographs. Please do not think that I am trying to discredit Pete, or the Miller, for the former was a great driver, and the latter a splendid car, but built for different conditions. Nevertheless facts are facts.

There are several other inaccuracies in "Wall Smacker" which indicate that Pete's memory was not very good. For example on page 68:

(1) He talks of his meeting with d'Annunzio, and of this poet and soothsayer, telling Chassagne to change the number of his Ballot from 9 to 11 as this would be the winning car number in the Gran Premio of Italy. As a result of this Chassagne is supposed to have done so and won the race. This is no so for the race was won by Goux (Ballot No.11) and Chassagne was second with Ballot.No.8.

(2) He stated also on page 68 that Sivocci was killed during the race. This is quite untrue for Sivocci continued to drive successfully until he met his end during practice for the I923 Italian Grand Prix driving the new P1 Alfa Romeo.

(3) On page 70 he says that O'Donnell and Gaston Chevrolet collided at Beverley Hills Speedway during the Thanksgiving Day race of 1921, both being killed in the resultant crash. This is untrue for it was during the same race but the year before!

As a result of these inaccuries I am a little wary of many of his statements, but perhaps he had an eye to the public for whom he was writing the book, and that a few exaggerations etc. were permissable as the normal reader would know little or nothing about automobile racing history. But, I repeat again I had a great admiration for him as a driver, particularly on board tracks, even if my opinion of him as an author is not very high.

It was a great pleasure for me to see you again the other evening, and I hope that we may meet again the next time you come to Paris.

With best wishes

(signed) Taso

Just nine days after the 1934 Indianapolis 500 there was a 100-mile race at Langhorne, Pa., which Scully, Philipps, and Moore were reluctant to miss. Lou had just given his all to win $5000 in prize money for the group, and that he should participate at Langhorne was out of the question. Joe Russo, an able driver who had just finished fifth in the Indianapolis 500, spoke up for the ride. Phillips recounted to Usher, "Russo was killed in that car at Langhorne. This turned Scully off of racing, and he quit the field for good." The Scully-Philipps partnership was wound up at that point, and in the settlement Bunny became the outright owner of the four-wheel drive mule. It's narrow 45-degree engine, if it ever could be repaired, was earmarked for his Type 35B Bugatti, which still was dear to his heart. He removed the battered power unit, put it on a shelf, and began getting rid of what was left of the car. "I gave the frame to Lou Fageol," he said, "and Lou Moore put the rear axle and transmission in the two-man car that Myron Stevens built for him."

With the partnership dissolved, Bunny was at liberty. Dick Loynes was in nearby Long Beach with heaps of Miller drawings, patterns, engines, and parts. "He really didn't know his way around that stuff," Bunny said, "so I went to work for him for a couple of years."

The organs of the cannibalized mule were dispersed, never to be reunited.

Chapter 14

Mauri Rose in the Driver's Seat

I 1935

By finishing second in the Indianapolis 500 in 1934, Mauri Rose became one of the hottest driving talents in the Western Hemisphere. He had worked hard to get there. He was born in Columbus, Ohio in 1905. He was intelligent, quick-witted, with little tolerance for the slowness of others. To his talent as a good mechanic and practical engineer was joined an intense competitive drive. He became a throttle jockey on dirt and boards in 1927, at age twenty-two, and he kept charging for twenty-four years, when he quit at age forty-six. Mauri got his first crack at Indianapolis in 1932, only to crash in practice. In 1933, he was the last of forty-two qualifiers to take the starting flag, but when his engine failed on lap 48 he had already hacked his way to third place. The 1934 race marked his real takeoff in the "big time"; he missed winning by less than two-tenths of a mile per hour. Lou Moore, who finished third and in his slipstream, would conceive and build the Blue Crown Specials that would give Mauri two of his Indy victories, as well as the rear-wheel drive Offy that would give him his third. But throughout the 1934 Indy experience the combination that apparently impressed Mauri the most was Brisko and the four-wheel drive. How he managed to

Illustration 14-1 In 1935, under Mauri Rose's management, the Four Wheel Drive Auto Company received a set of classical Hartford-type front shock absorbers (GBC ex IMS)

get that ride away from Brisko we may never know. There is almost nothing in what passes for the Company file for 1935, and that for 1936 is threadbare and strangely selective. Where did all that paper go, and why?

The year 1935 is represented by four documents. The first is a memo from Dodge to Olen concerning Rose as the new lessee of the Company's race car.

April 11, 1935

Mr. Olen:-

Regarding contract with Mr. Rose he has looked over the contract for 1934 with Mr. Brisko and the terms in this contract are satisfactory to him.

Mr. Rose has written out to two more drivers regarding securing their services as a relief driver and believe we will be successful in securing a satisfactory man to work with our mechanics during the month of May and thereby become thoroughly acquainted with our racecar.

Yours very truly,

H.B.Dodge
Chief Engineer

MIM

The next item is a practically verbatim copy of the previous year's contract with Brisko. Even the dates and the hours of the days are identical. One notes with a jolt that Rose is identified as being domiciled in Clintonville. This is not an error, and will be explained in the sparse correspondence of 1936.

THIS INDENTURE made and entered into this 12th day of April, 1935, between the Four Wheel Drive Auto Company of Clintonville Wisconsin, hereinafter called "Lessor" and Mauri Rose of Clintonville, Wisconsin, hereinafter called "Lessee."

WITNESSETH:

That the Lessor hereby leases to Lessee and Lessee hereby agrees to rent from the Lessor, one racing car described and knows as the "Four Wheel Drive Special" owned by the Four Wheel Drive Auto Company, for the term beginning May 16th, 1935 and ending on June 4, 1935 at 5 o'clock P.M.

It is expressly agreed and understood that said racing car may be driven by lessee in the 500 mile automobile race to be run at Indianapolis Speedway on May 30, 1935, to provide a competent and approved riding mechanic and also a competent, trained and approved driver in case one should be necessary as a substitute for the lessee, and to participate in said lease.

The lessor is to furnish all repair parts, lubricating oil, grease, gasoline, and tires required during the term of this lease, to put and keep said car in condition.

IN CONSIDERATION WHEREOF, the lessee binds himself to pay as rent for same a sum equal to 50% of the gross prize winnings or earnings of said car earned as a result the lessee participating in said race, said sum to be due and payable to lessor at the office of the Indianapolis Motor Speedway Corporation immediately upon payment of said prize money or earnings.

The lessee further agrees that immediately after the 500 mile automobile race on May 30,1935 at the Indianapolis Speedway, he will deliver said Four Wheel Drive Special car to the Four Wheel Drive Auto Company at Clintonville, Wisconsin, said delivery to be made at the point designated not later than 5 o'clock P.M. June 4, 1935, unless otherwise agreed upon in writing between lessee and lessor.

It is further agreed that lessor shall not be liable to said lessee for damages of any kind or from any cause whatever to the persons or proprty of lessee, or any person claiming by, through or under home and lessee agrees to defend at his own expense all damage suits which may be brought upon claims relating to said cause or the use thereof by the lessee.

IN WITNESS WHEREOF, the parties hereto have hereunto set their hands and seals the day and year first above written.

Attest:	FOUR WHEEL DRIVE AUTO CO. Lessor
Frank Gause, Secy	Walter A. Olen, President
Witness: H.V. Larson	Mauri Rose Lessee

Then comes a half-page clipping from an unidentified and undated newspaper. Most of the space is devoted to a pre-Indy feature on the Miller Fords, and there is a photo of Harry Miller and Preston Tucker watching work in progress. The clipping lacks any relevance to this file. There is no sign of any pre-race publicity, nor of race coverage. It is as though no one cared, or the file was gutted.

Contemporary news coverage of the 1935 Indianapolis 500 tended to concentrate on the Miller Ford rout and Kelly Petillo's win with a four-banger called an "Offy." As for the performance of the four-wheel drive, Rose qualified the car in tenth position. He drove a beautifully steady race, having methodically moved up to eighth at fifty miles, seventh at one-hundred, fifth at one-twenty-five, and third at one-seventy-five miles. Had he been able to keep going, he well might have won, but he was put out by something identified as a "broken water manifold stud." Mauri and the car were just also-rans.

Illustration 14-2 Although Mauri Rose signed this photo in 1936, the lettering on the car frame is that of 1935. Here he rests a hand on the lever of one of the new shocks. The transaxle housing also contained reservoirs for water and oil. (RMRC)

The only other document we have that refers to 1935 is this bit of tire data. Waldo Stein is memorable for having inherited Miller's Golden Sub from Barney Oldfield.

Illustration 14-3 In 1935, pipe smoking Rose qualified tenth at 116.470 mph. Driving beautifully, he was in third place when some vague mechanical failure put him out of the race. (RMRC)

June 10, 1935.

Mr. Olen

Mr. Dodge

Mr. Hanson

We made an accurate check on gasoline mileage of FWD race car Decoration Day on the first seventy-one laps when car was called in for gasoline. This mileage also includes two five minute warm up periods before the race.

Miles per gallon, average, for the first 177 miles was - 14.8.

The average speed of this period was brought down to 108 miles per hour due to accidents. The majority of these laps were run at an average speed of 111 miles per hour.

Highest speed on any one lap - 114.5 miles per hour.

In checking the wear on the tires the wear seems to be more uniform this year than last indicating better alignment. Waldon Stein of the Firestone Company also checked these tires, and stated that he would write us and give us official figures on comparing our tire wear with some of the best rear drive race cars. He was also very much surprised at the slight and equal wear of our tires this year.

Yours very truly,

H.V. Larson

H.V. Larson
Experimental Engineer.

MIM

II The 1936 Season

This year's file contains all of eighteen pages, instead of only five. One odd fact is that almost half of those belong to letters written by Rose and addressed to Larson, his official contact in Clintonville. The correspondence is one-sided; Larson's replies are not preserved. There is a psychodrama going on, in which Mauri's situation becomes increasingly poignant. His first letter is that of a self-assured professional who knows precisely what he wants and needs. He is clearly domiciled in Indianapolis and his relationship with Speedway Motor Service, aside from having access to its letterhead, never is defined. Existing transfer-case oil lines must go.

Jan 17th, 1936

Mr.H.V.Larson
F.W.D.Auto Co.
Clintonville,Wis.

Dear Herman:

Just received your letter of Jan,16, and will answer your queries, so you will know what's what: Regarding the raduis rods, they will not lighten the car any, but, will keep the rear axle in better alignment, giving better mileage and speed. Also increase the safety of the car.

As far as the carburetion is concerned, we will have to get as much more economy as we can, so we can run faster than we did, or than anybody else can run. There is two way's to do it. Either different carburetion and manifolding, or, go to a little higher gear. The 3.09 is a little too high, the 3.25 a little too low, for the economy that we want. A 3.15 gear would be ideal, I believe. Dont make the mistake of thinking that a 3.15 gear would'nt make much difference. A higher gear is needed because of the changes in the track. It is not necessary to slow down as much in the turn now, as it is smooth, and you can keep the R.P.M.'s up. With a 3.15 gear we can get 15 miles to the gallon.

The pistons will have to be machined here, as you have no equipment at the factory to do the job. The pistons have to be relieved at the top, above the rings, remember where the piston that stuck scored, above the rings. I have seen everybody else's pistons in 255's and they are relieved above the rings. We can get them machined at the same place as we did last year, The Century Tire Co. here in Indianapolis.

As for the magneto, I expect an answer from them soon, and will give you the name and no. magneto to order. This magneto does not rotate any winding, is lighter, and more reliable than the Bosch. It is the type C, Spitdorf magneto. I wrote them for all the particulars, and will forward the information to you just as soon as I receive it.

What have you done in regard to the oiling system of the transfer case? Unless you are willing to make these changes in the car, I will not be interested in driving the F.W.D. So, make up your mind, and let's get going, one way or another.

Yours very truly,

Mauri Rose.
902-N.Pennsylvania St. Apt. No.104

It was the dead of winter, with no racing going on and no money coming in. Mauri felt that he could be useful in Clintonville, rather than sitting on his hands in Indianapolis.

Jan.23rd,1936

Herman Larson
C/O W.D.Auto Co.
Clintonville,Wis.

Dear Herman:

Just received your letter and will answer promptly so you will know that everything is O.K. with me as far as the F.W.D. is concerned. I prefer driving the four-wheeler to any other car on the track, except that it is so hard to convince you that there are some changes that have to be made, because the track is different. Remember the trouble I had trying to convince you that it was the magneto that was bad, last year? Well, let's forget all that and start out anew this year:

I will agree with you that rear radius rods are not necessary, and would add weight, etc. So we can skip them. I am glad to hear that you have taken care of the transfer case and oil lines I would recommend that you order the 3.165 gear ratio, right away so we will be sure of getting it, and also give us plenty of time to install these gears. I am sure that this 3.165 ratio will give us even better mileage, and performance than we had last year.

I am sure that we can work out the rest of the parts such as pistons, etc. I have the latest information on the clearances and method of machining them. I will send you the dope on this Splitdorf magneto just as soon as their engineer sends it to me. The short wire to the steering wheel has a lot to do with the performance of the magneto, and they can furnish us a magneto with a built-in switch operated by a control wire. The way it is now, the short wire acts as a shorted condenser on the magneto current.

I will be very well satisfied with the set-up now, except that I am not doing anything down here, and would like to come up to the factory and work on the car, or anything else that I could do up there. It should keep me busy on the car, if there is bearings, gears, and the motor to go over, besides the rest of the chassis. I do not think this is out of line, as I would like to see the car go together myself, and there is a lot I can do, as I have had the experience. I feel sure that we can win this year, and hoping that arrangements for me to come up and work can be made satisfactorily. With best regards, and looking for an early reply, I remain,

Yours very truly,

Mauri Rose

Mauri Rose.
Apt.No.104
902-N.Pennsylvania St.

The dread transfer case still was not entirely tamed. But Mauri's confidence was total. The Century Tire Co. belonged to the family of Karl Kizer, the father of the Speedway's Hall of Fame Museum.

Feb. 12th, 1936

Mr. Herman Larson
F.W.D. Auto Co.
Clintonville, Wis.

Dear Herman:

Just received your letter and will answer promptly so you will know that everything is O.K. Glad to know that you have taken care of the transfer case and lines. Could you make the lines so that they will not be in the way of the "accelerator" leg. This improvement would make the car more comfortable for me. You may rest assured that I am willing to go more than 50-50 with you on this years race. In fact, we both know that we have a very good chance, at first place this coming race.

The clearance for the pistons is .012" and cam ground, with a slight taper from bottom to top. The "Century Tire Co." can machine these pistons for us accurately, and with a minimum of expense to the Co. You can send them to me anytime and I will get them started on the pistons and get them out while they are not so busy. We had better have all five pistons machined, so we will have a spare piston, just in case we need it.

I have the number assigned to me for this year at the "500" race, so if you paint the car, you will know the no. to put on the car. It is no. (36} Thirty-six. I have no preference, or suggestions to offer for painting the car. You will be able to have the same garage, if you prefer it to any other. Will take up the magneto question when I see you at the factory. Will be glad to work with you on anything that comes up, or you are not satisfied with.

Hoping that this takes care of everything up to date, and hoping to hear from you in the near future, with best regards and my assurance that I will do my part. I remain,

Yours very truly,

Mauri Rose

Mauri Rose.

P.S. Tell Dodge not to worry about a relief driver, I could go "1000" miles in the "Four-Wheeler", if necessary, with-out relief. All we have to do is put the car in good shape. I'll take care of the rest of it.

The Company apparently was ignoring Mauri's proposals concerning employment. In frustrated confusion, he addressed this three-page soliloquy to Larson.

Feb.19th,1936

Mr. Herman Larson
F.W.D. Auto Co.
Clintonville,Wis.

Dear Herman:

Just a line to inquire about the old V-8 motor that the company has for sale. I have someone interested in buying it to use in a boat, which is the only place it can be used. It is not worth ten cents, as a race car engine is concerned. The company may not know that this is so, but you and I do.

If they want to sell it at a very reasonable figure I may be able to promote the sale of this engine for the company. It will have to be a very low price, for this fellow to buy it, as he will only use it for pleasure in a boat. Please get me the low-down cash figure for this engine, and I will see what I can do.

What is new up at the factory? Have you been able to see Mr. Olen about me coming up in the near future, to work on the car. If he is going to wait until in April to decide, I don't believe it would be worthwhile for me to come up, as, I have a good one man car to drive this season, that will both run and stay together. And I can get some pretty good appearance deals to race in the East, the latter part of April. There is also a well paying race in Langhorne Pa. early in May.

I would prefer coming up there the first of March and working until the middle of April. We could have the car all ready by that time, to bring down here, and get ready for the "500". After all, Herman, I make my living by driving race cars; and if I have a chance to make four or five hundred dollars in appearances, before the "500" I would be foolish to pass it up, unless I can benefit in another way, like working for them (factory) now while I am doing nothing.

In all probability, the only value that I have to the F.W.D. factory, is, that I can drive their race car, and maybe win the "500" mile race for them. That, chiefly, is what they are interested in doing, or they would-never have contacted me in the first place. I am willing to do anything that is fair, but, I have to make a living, too. So, you see, I am in a spot: I would like to have your opinion, also Mr. Olen's and Mr. Dodge's on this matter.

When I went to work for the F.W.D. last year, it was with the understanding that I receive $150.00 per mo. and expenses. I can truthfully say that, I did not receive any expenses, which I incurred moving up to Clintonville, and keeping up a residence there while on the road for the company, which was quite an expense, also the fact that I did not "chisel" on any of my expense account while on the road for them.

In fact, I did not make any money or get rich on the co. in any way, and was honest enough to not take a pay-check after-ward to work in the sales dept. because I knew that I was not worth that much salary to them in the sales end. Possibly I made a big mistake in not working in the sales dept. and take a check every month for doing nothing. Since I have started on this, I may just as well say all I have to say and so you will know the whole story.

I would like to know just how Mr. Olen feels toward me and what plans, if any he has in mind for the race car and myself. Let me state again, that I prefer driving the F.W. Drive car in this coming race, to any other, but, look at this from my side once. I am a race driver, and there is only two things that I can do. That is, either drive all the races I can, make as much money as possible, and when I am no longer able to drive a race car, go into some business for my-self, so I will be able to earn a living the rest of my life, or, while still able to drive race cars, and, if I get a break and place well up or win the "Indianapolis" race, I get a position with some large company, where I can work into a successful future, in the line of work that I know I can do successfully, quit racing and settle down to this job, and do my best to make something out of myself.

If I don't do this before I am thirty-five, or so, I will never get anywhere. So you can see that what I do this year really means more to me than, just another race. And, just what I am up against. I would like to obtain a position with a company, doing the work that I am capable of, and that I like, that I will have my heart in there with me, pulling. Like I do driving a race car. I like it, my heart is in my driving, and I can do it well. But; there is no future in just driving a race car. I am only good for so many more years of it They are numbered in racing, just as in prize fighting or any-thing else.

Of course, I can go on racing until I am "punch-drunk", and useless, but, this is not my aim or desire. In other words, I am at a marked crossroad of life, one goes to nothing; the other two?? Nobody knows. Is my future with the "Four-Wheel-Drive Auto Co."? or with some other concern in the automotive field? I have sense enough to know what a future planned on "racing" would be like. But, I do not know, where the other roads will lead, but, I have to find out, for sure, before I start out on one of them, as these roads are narrow, and there is no backing up, or turning around.(stop, fill up pipe)

And, even if I could back-up, or turn around, after I found out that it was the wrong road; when I got back to the cross-roads, I would be just where I started, and "thirty-years too late" to do any thing about going on another road. Where this all leads to, I am not even sure myself, but I would like to be before I start one way or another,(stop, fill up pipe) .

I will say this; I am going out this year and try to win this "500" mile race, and I would like to do this with the "Four-Wheel Drive" race car. To do this, we know that the car must be put in the best shape it is possible to do. All we are interested in now, is, the best and easiest way to do this. If Mr. Olen and Mr .Dodge, or your-self, think that it would be better, or more advisable, to do the necessary work on the car, yourselves, at the factory with-out my help it will be perfectly O K.with me, and no hard feelings. And I would be just as willing to drive it, as if I came up to the factory and worked on the car.

If I can be helpful in putting the car in shape, I would like to come up to the factory, but, if they do not think so, I prefer to stay here. I do not want you up at the factory to think that I want a soft job, and salary, to keep me going. In fact, I would rather race, because, I can make more money racing before the"500" than I can working at the factory on the car, or other work up there. The point is, whether I can help put the car in better shape. It is not whether I am going to make, what they believe to be an excessive salary.

I will be here at the track anyhow, and can do anything to help on the car then(after you bring the car down)so all I want you to do, is, the best thing for both, the company and myself. But, if I do not work for you up there, I am going to do some racing the latter part of April, and early in May. Possibly this will be the best thing to do, as, I will really only work to the benefit of the company, on the day of the race and during the race, and I assure you that I will really turn on that day, May 30th.(stop, fill up pipe)

At any rate, will be glad to hear from you at any time, and again assure you of my complete co-operation with you on this race at Indianapolis, May 30th. With my best regards to you, and the company, I remain,

Yours very truly,

Mauri Rose

Mauri Rose
Apt.No104
902-N.Pennsylvania St.
Indianapolis,Indiana

Mauri's repeated proposals to be taken back on the Company payroll apparently continued to be ignored. At last he tried spelling out his own terms.

March 4th, 1936

Mr.Herman Larson
F.W.D.Auto Co.
Clintonville,Wis.

Dear Herman:

Just received your letter of March 2nd,and will answer promptly so you will kmow that everything is alright. I also received the shipment of pistons today. Will get busy on them right away. I saw Mr. Merz to-night and he said the best thing for us to do would be to take the front and rear axles, spindles, and steering arms to the "General Electric X-Ray Laboratory" in Chicago, Ill. and have them X-rayed, and get a dated certificate from them. This would be approved by the technical A.A.A. board.

If we have to tear the car down to have this done, it will take a little time to do this, and put it back to-gether again. If we wait until we get down here to do this work, it may be delayed quite a bit, as everybody else will want theirs first, and I am not sure that they will be able to take care of the work here in the first place. Merz reccomended that we do this X-ray work before we brought the car down here. I think that that is the thing to do, myself.

As I said before I would like to see the car go to-gether myself, and there are several way's to look at this. I really want the car in A-1 shape for this race. The way things look now there would be enough to keep me busy the month of April on the race car. I am working now, but would quit the last of this month, and come up to Clintonville and try to put the car in good shape for the race, as it means more to me than the job I am on now. As for after the race, we can work out something then, I'm sure.

I may not make as much money working at the factory on the race car, as driving the early races, but, it probably would be safer, and the smartest thing for me to do, as I am sure I don't want to miss driving the F.W.D. in this coming "500". If it is satisfactory to Mr. Olen and Mr. Dodge that I come up to work April 1st, at $150. and expenses, per month, I will make arrangements to be in Clintonville by April 1st, ready to go to work on the car.

Will be glad to have "Arnie" work with us again this year, he is a good man to have with us on the race car. I will have a car to use while we are here so that part is O.K. I have already made arrangements for fuel (Gilmore)and either mineral or castor oil, which-ever you prefer to use.

We can make all other arrngements when I see you, and if there is anything you may want from here, or that is not satisfactory to you, please let me know and I will do my best to do what-ever you want, or require.

Hoping that this clears everything up, and to hear from you in the near future as to my coming up to Clintonville to work on the car. Will close with best wishes and regards, I remain,

Yours very truly,

Mauri Rose

Mauri Rose
902-N.Pennsylvania St.
Apt.No.104

After the passage of another five weeks the now-standard lease contract was retyped and signed, this time by Olen himself. Once again Rose appears as an inhabitant of Clintonville, suggesting that perhaps at this juncture he had been allowed to aid with the car's preparation at the factory.

THIS INDENTURE made and entered into this 11th. day of April, 1936, between the Four Wheel Drive Auto Company of Clintonville, Wisconsin, hereinafter called "Lessor" and Mauri Rose of Clintonville, Wisconsin, hereinafter called "Lessee".

WITNESSETH:

That the Lessor hereby leases to Lessee and Lessee hereby agrees to rent from the Lessor, one racing car described and known as the "Four Wheel Drive Special" owned by the Four Wheel Drive Auto Company, for the term beginning May 16th, 1936 and ending on June 4, 1936 at 5 o'clock P.M.

It is expressly agreed and understood that said racing car may be driven by lessee in the 500 mile automobile race to be run at Indianapolis Speedway on May 30, 1936, to provide a competent and approved riding mechanic and also a competent, trained and approved driver in case one should be necessary as a substitute for the lessee, and to participate in said lease.

The lessor is to furnish all repair parts, lubricating oil, grease, gasoline and tires required during the term of this lease, to put and keep said car in condition.

IN CONSIDERATION WHEREOF, the lessee binds himself to pay as rent for same a sum equal to 50% of the gross prize winnings or earnings of said car earned as a result of the lessee participating to said race, said sum to be due and payable to lessor at the office of the Indianapolis Motor Speedway Corporation immediately upon payment of said prize money or earnings.

The lessee further agrees that immediately after the 500 mile automobile race on May 30, 1936 at the Indianapolis Speedway, he will deliver said Four Wheel Drive Special car to the Four Wheel Drive Auto Company at Clintonville, Wisconsin, said delivery to be made at the point designated not later than 5 o'clock P.M. June 4, 1936, unless otherwise agreed upon in writing between lessee and lessor.

It is further agreed that lessor shall not be liable to said lessee for damages of any kind or from any cause whatever to the persons or property of lessee, or any person claiming by, through or under him, and lessee agrees to defend at his own expense all damage suits which may be brought upon claims relating to said cause or the use thereof by the lessee.

IN WITNESS WHEREOF, the parties hereto have hereunto set their hands and seals the day and year first above written.

FOUR WHEEL DRIVE AUTO CO.
Lessor
BY (signed) Walter A. Olen,President

Attest:
(signed) Frank Gause

Lessee
(signed) Mauri Rose

Witness:
(signed) Helen Neller

On May 8th, 1936, Larson in Indianapolis wrote to Dodge in Clintonville:

Dodge

Called Ross Gear & Tool Co this morning and ordered two steering gear arms. We have an extra arm on hand that is 8" long. It also looks bad and has not been approved yet. We will not have any trouble in getting a good Steering Gear arm from Ross.

We did not get the steering arms yet, but expect to get them late tonight at the express office, as soon as we receive these forgings Merz will have somebody work Saturday and Sunday. They have the equipment for this kind of work.

Be sure to send me the material specifications for the forgings, Sorensen has these.

Have been working on the old steering arm filing out the defect have most of it out, but don't think they will pass it. Will try Saturday they have been awful busy today rejecting spindles and arms. One of the Boyle front wheel drive both spindles and arms were rejected. Louis Meyer had both his spindles rejected today.

Yours truly

H.V. Larson

H. V. Larson

That this next scrap of paper has come down to us is a marvel. It is a copy of a telegram sent to Dodge thirteen days before the race by Company man E.M. McLean in Los Angeles—he who used to visit the Miller plant when the car was being built:

COPY

GU13 51 COLLECT WA HOLLYWOOD CALIF MAY 16 1157A
H B DODGE

FENGLER OWNS OTHER CAR WILL RENT FRONT AXLE FOR RACE FOR ONE HUNDRED TWENTY FIVE DOLLARS OR WILL RENT SAME AND GIVE YOU COMPLETE NEW TYPE TRANSFER CASE AND TAKE V EIGHT MOTOR AS PAYMENT PROVIDING MOTOR IN GOOD CONDITION STOP WIRE INSTRUCTIONS CARE BRANCH STOP NO OTHER AXLE AVAILABLE

E M MCLEAN
229P

Illustration 14-4 The accident deducible from this telegram left the car's front end badly mangled. It took place during practice about two weeks before the race and launched a panic quest for the pieces of the cannibalized mule. (RMRC)

For such a short message, it contains a wealth of information, from which it may be deduced that:

1. Mauri Rose had had an accident and had bent the rather delicate front end of his car. Or, its front De Dion tube had been rejected by the technical committee. Or both.

2. The only source in the world for the damaged part(s) was Harlan Fengler, then teamed with Lou Moore. According to historian Bill Digney, they had already recycled the rear axle and suspension of the four-wheel drive mule in building Moore's Burd Piston Ring Special. (Powered by a Miller 255, it was driven by Fred Frame in the 1936 Indianapolis 500, retired early. With a new Offy 270, Mauri Rose lasted for 127 laps in 1937 before an oil line broke. In 1938, the car won the big race, Floyd Roberts up.) Brisko would get the front end.

3. Independently of the Four Wheel Drive Auto Company, and presumably by the old Miller crew, a transfer case had been designed and built which avoided the vices inherent in the original. Thus, De Paolo was able to go the distance at Tripoli in perfect serenity. Whatever became of this new transfer case?

4. Not everyone was writing off the Miller V-8 engine.

To check out point No. 1, above, one had only to consult the contemporary race coverage. Exact! Mauri had smacked the wall in practice again this year.

This scenario envisages Fengler's axle and accoutrements being flown to Indianapolis, making possible the following note, dictated to Dodge by Larson on May 22nd: "Got race car together and out on track at 3:30 today. Mauri drove 4 laps at about 30 m.p.h. and then one fast lap. Then drove in. Adjustments must be made before qualification trial. Will work on car tonite."

With three days to go before the race, Dodge managed to remain calm in reporting to Olen that Mauri considered the car to be "mechanically perfect." The borrowed carburetors were taken away, and could not be replaced and tuned until the following afternoon. With this telegram, dated May 26th, the 1936 file comes to a premature and hollow end, once again as though nothing had happened and no one could have cared less. Instead, Mauri Rose outdid himself to put the FWD Special through its finest racing performance ever.

PA 4 62 COLLECT DL SW INDIANAPOLIS IND 320PM MAY 26 1936

W A OLEN

CLINTONVILLE WIS=

BRIEF TEST YESTERDAY HAS CONVINCED MAURI THAT CAR IS PERFECT MECHANICALLY STOP CARBURETORS USED YESTERDAY NOT AVAILABLE TODAY DUPLICATES BEING SUPPLIED BY WINFIELD PERSONALLY THE MAKER OF SAME BUT NOT AVAILABLE BEFORE LATE THIS EVENING STOP MAURI THINKS UNWISE

TO MAKE TEST TODAY BUT WILL BE ON TRACK BY TEN TOMORROW AND ATTEMPT TO QUALIFY SHORTLY AFTER FOUR TOMORROW CLINTONVILLE TIME =

H B DODGE
344PM

Illustration 14-5 Somehow the car was gotten back together sufficiently for it to barely qualify for the race. Clintonville brass understandably did not stay around for such show as there might be. However, man, crew, and car performed wonders, Mauri finishing the 500 in fourth place. (RMRC)

What happened was that Mauri finally managed to qualify a very sick car in thirtieth place, out of a field of thirty-three. His average speed was 113.890 mph, almost as slow as McDonough with the same car in 1932, at 113.276 mph. Since then Brisko, not a superfast driver, had averaged 118.388 mph in qualifying the machine. But while the Clintonville contingent was heading for home in disgusted wrath, Mauri and his crew sweated over the car. We don't know what they did, but, if one can believe the coverage in *Motor*, Mauri had blasted his way to fourth in the first lap of the 500. If true, the team in his pit revolted and he backed off. He went back to his familiar, steady style, arriving in eighth position at seventy laps, sixth at a hundred and forty, and fourth at a hundred and fifty. From there until lap two hundred he stroked along between fifth and fourth, finishing in the latter position 1.8 mph slower than winner Lou Meyer, whose obviously healthier rear-wheel drive Myron Stevens chassis also was powered by a Miller 255. The only evidence we have that indicates that any of this was noticed in Clintonville is inclusion in the file of a Champion spark plug advertisement that celebrates Mauri's winning of the A.A.A. National Championship that year. The car pictured in the ad is the FWD Special, which he drove only at Indianapolis. It had not been a convivial two-year relationship.

Illustration 14-6 Champion advertisement featuring FWD special.

Chapter 15

Pete De Paolo, Racing Impresario, and the Final Racing Season

This year's file begins with a reply from Tripoli and Avus veteran De Paolo to a letter, lamentably missing, to him from Olen. This in itself is an unexpected event: Olen, traditionally the recipient of proposals, is initiating one. And what a one! "A campaign for at least a year, both in this country and in Europe, with a team of four-wheel driven cars." And here we encounter Miller's four-wheel drive mule again, "scattered all over" Los Angeles. Olen apparently wanted to check the feasibility of putting it back together again. It would be wonderful to know more, such as where the money would come from. And just what Olen and his associates thought they might do against the German steamroller in Europe, using unblown four-bangers in 1932 chassis. Who was hallucinating?

March 28th,1937

Mr.W.A.Olen,President
The Four Wheel Drive Auto Co.
Clintonville,Wis.

Dear Mr.Olen:

I have just returned from a trip in the Northwest, in behalf of the company I now represent, and will until May 15th, and I find your kind letter of March 17th, which is very interesting indeed.

Tomorrow I am holding a meeting in Oakland, and Tuesday a meeting here in San Francisco, then I will go on back to Los Angeles, where I will stay until about April 15th, then will start for the east, I will keep you posted as to my whereabouts, so that we might be able to get together on a deal of some kind that will be satisfactory to all parties concerned.

In regards to a proposition such as you mention in your letter, a plan whereby I would manage your car, or cars, and I to provide the drivers and mechanics, sounds very good to the writer, and I am sure we could garner a great deal of advertising material in this manner, in behalf of The Four Wheel Drive cars, and between races we could arrange various meetings for me to conduct throughout the country, as well as Europe.

As quickly as I get back to Los Angeles, Wednesday or Thursday, I will first of all try to get in touch with your sales Mgr. Mr.R.H.Schmidt, then get all the information I can regarding the Four Wheel Drive car that is scattered all over town, and, learn for certain whether or not Mauri Rose has signed up with Lou Moore, as Moore advised me prior to my departure from Los

Angeles that Mauri would drive for him at Indianapolis, and the balance of the year. I know Muri very well, and I am sure that if we entered into any kind of a deal, whereby I would manage your car, or cars, he would gladly consider my proposition with him.

As for the problem of a reliable mechanic on the car, just leave that to me, I will get the best, either Ernie Olsen or Jean Marcenac, two of the best boys in racing, both knowing Four Wheel Drive cars, and both having worked for me in the past.

As quickly as I get back to Los Angeles, and get all the information, I will write you air-mail accordingly, and in the meanwhile, rest assured that I am terribly interested in your idea of having a campaign for at least a year, both in this country and in Europe, with a team of four-wheel driven cars.

I am writing my publishing company at this time, instructing them to ship to you twenty-four (24) copies of my book "Wall Smacker", and I am grateful for your kind order.

With kind regards, I am

Sincerely yours

Peter De Paolo
4225 1/2 Avocado St.
Los Angeles,Calif.

This file contains nine letters from De Paolo to Olen, most of them lengthy. At a great loss to posterity, all of Olen's letters to De Paolo are missing, with the exception of the document that closes the file and this whole era. It suggests that, throughout this relationship, Olen was charmed by De Paolo in the same admiring and warm way that he had been won by Oldfield. We learn here that, within ten days of De Paolo's original reply to Olen, the latter had committed his company to a new attack on Indianapolis, with De Paolo in charge. De Paolo's homespun humor and cracker-barrel philosophy seemed to resound perfectly in Clintonville.

Los Angeles,Calif.

April 12th,1937

Mr.W.A.Olen,President
Four Wheel Drive Auto Company
Clintonville,Wisconsin

Dear Mr.Olen:

In answer to your kind wire and letter of the 7th, it is needless for me to mention how happy I am now that our deal has been completed, for the necessary preparations of an Indianapolis "Winner", and you can rest assured I will do all in my power to put over another victory, this time in behalf of your company.

I am terribly sorry that I will not be here to meet Mr.Schmidt, and have a visit with him, as I must leave tonight by air for Waco, Texas, my meeting is scheduled there for tomorrow night, then I will go on to Houston, New Orleans, Birmingham, Atlanta, etc., as quickly as I get a definite itinerary from the company I will send it on to you, and Mr. Schmidt, in order that we may keep in touch with each other, as related to you in previous correspondence I will continue with my present position until May 15th,after which time I will be in Indianapolis, and in behalf of your company until after race time, and thereafter if necessary arrangements can be made along those lines, which we will discuss when I see you personally.

I have instructed Jean Marcenac to be in Indianapolis, May 1st, with instructions to go through the car thoroughly from stem to stern in accordance with his satisfaction, I will keep in touch with him constantly until my arrival there in Indianapolis May 15th, and we can go into our old time huddles, throwing the "Wop" and French lingo into the car, and then combine our efforts with our "Swede" driver, and we'll have the other fellows baffled, truly a cosmopolitan team, harmony in our team will prevail, which is very essential for victory, and in accordance with my old rule, we will "plan our work" and then "Work our plan", just can't miss..

I will keep in touch with you Mr.Olen, until I arrive in Indianapolis, then we can make other arrangements that you might desire to make, in the way of your coming to Indianapolis race day etc..

With kind regards

Sincerely yours

Peter De Paolo

It would seem that in his original letter to De Paolo, Olen had suggested using Mauri Rose as driver again. Here it becomes clear that all contact with Rose had been lost by the Company, in spite of Rose's fine performance against very bad odds in 1936. De Paolo, who had not had a good ride at Indianapolis since 1926, had begun to fantasize about a comeback.

April 20th,1937

Mr.W.A.Olen,President
Four Wheel Drive Co.
Clintonville,Wis.

Dear Mr.Olen:

I received a telegram from Jean Marcenac, my mechanic last night saying he was leaving for Clintonville Wednesday, April 21st, and no doubt will report there to you on Monday. I am sure you will enjoy meeting Jean, he is very interesting, and truly the best mechanic in the business, along with being a real student in whatever endeavor he might undertake, therefore I have instructed him to learn all he can while there at your plant.

Jean also informed me in his letter that you were trying to get hold of Mauri Rose, if you can locate him I feel sure he has signed up with Lou Moore, if this is not true, I would like very much to have Mauri as the driver as he is well acquainted with the car, and with Jean and myself behind him I know he will do a much better job.

I am trying to arrange to be in Indianapolis not later than May 10th,for as I have already written you my present deal does not expire until May 15th, however I have written to the company requesting I get back to Indianapolis by May 10th, and I feel sure they will comply with my request..

Yesterday I received a letter from Miss Dallenbach, from the Indianapolis Speedway office, and I am enclosing her letter which is self explanatory, and if your office has not mailed the duplicate of the entry blank to Washington, please do so, or send it to Miss Dallenbach to Indianapolis.

In my letter to the Speedway with the signed entry blank, I requested they shoot all they possibly can in behalf of the Four Wheel Drive Company, in connection with my name, and I know they will do a good job, which might come in handy for the future with us.

In regards to the driving of the car at Indianapolis, it is my thought to try the car myself when I get there, and if I still have the "old touch" to still drive in a winning fashion as of old, our driver situation is completely settled, however I do not want to kid myself, for after all I have been out of it for almost three years, and I wouldn't undertake the job unless I could do justice to the merits of the car, for I have my mind on just one thing, and that is "Victory" for the Four-Wheel-Drive car, whether I drive, or manage someone else.

I will be in Philadelphia from Saturday, April 24th,until April 27th, then I will go to New York.

April 24th--27th-- Phila. Hotel Benjamin Franklin
April 27th--31rst-New York City--Lexington Hotel

After this date I will notify you as to where I will be, roughly I can say it will be towards Indianapolis, Rochester, Buffalo, Cleveland, Detroit, Indianapolis, ... I will advise you more definitely from New York next week.

It would be impossible for me to completely express myself in this letter, how thrilled I am on having your car in my command for the coming race, and you can rest assured Mr. Olen that it will get the very best attention, towards preparing it for the race, and after the victory--oh boy!!!! we will tell the whole world about it..

With best personal wishes,I am

Very sincerely yours

Peter De Paolo

Peter De Paolo
P.S. Enclosure

The letter that Pete De Paolo enclosed with the foregoing one to Olen attested to his being on the job and a part of the Indianapolis Motor Speedway family.

April 15, 1937

Mr. Peter DePaolo
c/o Hotel Bankhead
Birmingham, Alabama

Dear Peter:

It sure was grand to receive your airmail letter this morning entering car for the race on May 31st. It seems like old times.

Have noted carefully all you have to say and have relayed same to the Publicity Department. Will give you as much of a break as possible.

It will be nice to see Jean again as we have missed him the last few years. Evidently, you detached the duplicate for the AAA from the entry blank as it was not enclosed with the rest of the blank. Did you send this to Washington? If you still have it, it would be a good idea to send it either to this office or to Ted.

Best regards from the "gang".

Very truly yours,
INDIANAPOLIS MOTOR SPEEDWAY CORPORATION
(signed) E.S. Dallenbach
Secretary to T. E. Myers
ESD:MM

From De Paolo to Olen's secretary, the usual gratuitous, excessive verbiage:

April 26th,1937

E.L.Vaughn
Secretary to the Pres.
Four Wheel Drive Auto Co.
Clintonville,Wis.

Dear Mr.Vaughn:-

I am extremely grateful for your kind letter of April 23rd,with the duplicate copy of the Indianapolis entry blank enclosed therein, I will send this out immediately to the AAA in Washington,D.C..

No doubt Jean Marcenac has arrived in Clintonville by this writing, and you will have made the necessary arrangements for him to get the xx car finished to get started for Indianapolis.

I am enclosing herein some newspaper clippings which I thought might interest Mr.Olen, when he is finished with them please pass xxx them on to your advertising (please excuse the mistakes, my own "stenog" this morning) dept., and assure this department I will be shooting in a lot more.

Looking forward to the pleasure of a visit to Clintonville in the near future, and have the good fortune of meeting you all personally, I am

Sincerely yours

Peter De Paolo

Peter De Paolo

P.S. At the above address until Friday this week, then I will be at the Lexington Hotel, New York City, the following week. P.

Pete de Paolo walked in and sat down.

The Italian kid who was the speed king of the world in 1925 and again in 1927, is trying hard to be a businessman now—selling piston rings, or, rather, promoting them.

He toyed with the idea of driving in the Iron Derby again this year. He thought of it. And he thought of it from every angle.

Time was when he would tie a pair of his baby's shoes to the frame of his car and put his foot down hard—fighting the 500 miles of bouncing brick at Indianapolis—the board tracks and the dirt tracks.

PETE DE PAOLO.

But the bambino no longer is a bambino. And Pete de Paolo no longer is living in 1927 or 1925.

"It's 1937, isn't it?" he asked.

He might make a great race of it. In fact, I think he would. But, after all, the game belongs to the kids. And not many men have twice been champion of the world.

So, when the motors break into their stuttering roar on May 30, and gradually warm up to a steady, pulsing inferno of sound . . . and when the stench of burned castor oil begins to float on the air . . . Pete de Paolo will be a manager in one of the pits He'll manage a four-wheel drive car with a Miller motor.

"You come up and sit in my pit," he said. "You can ride in one of the practice turns—"

There I stopped him. I did that once. That was in 1927 And just once was enough. The breath goes. And the stomach and the other internal organs seem to pull out of the body on the turns. And we were not going so fast—as I recall just 90 miles an hour. But today they average more than 100 miles an hour for the entire 500 miles.'

"Well, anyhow, you come and sit in the pit. That will be a new story.'

The "Iron Derby"—500 miles of speed and 150,000 or 200,000 looking on—America's greatest sporting spectacle. You can't argue any other way. The crowd proves it.

The craze for speed—and some say the morbid pull of the possibility of a wreck—with the scream of the meat wagon's siren going after some poor devil who has gone over the rim

Continued on Second Sports Page.

BREAK O' DAY

Continued From First Sports Page.

or crashed on a turn—well, it's more than that. But it's all a part of the picture.

AN INTERNATIONAL INCIDENT.

Pete de Paolo was driving a four-wheel, special-motored car in a series of races in North Africa, Germany and France a few years ago.

And he almost brought about an international crisis.

He was driving in a race at Berlin on the Avus speedway.

"I was doing swell," said Pete, "until it started raining torrents. I was still doing swell after I got used to the rain.

"But all of a sudden two connecting rods broke and the engine came apart. One of the rods almost hit Adolf Hitler, who was in the grandstand.

"I was lucky to get out alive. Hitler was very nice about it," said Pete.

The little Italian has been in lots of crashes. He has been picked up for dead. And yet, here he is, not a scar showing and going strong.

A BIT OF GOLF.

"Any talk hereabouts about John (Mysterious) Montague?" asked Pete de Paolo.

"Plenty."

"Well, I know him well. And my brother, Johnny Paolo, who is a pretty fair golfer himself, is a close friend to Montague.

"He's all they say he is. He has the golf world in a stew. No one knows from whence he comes. No one knows where his money comes from. No one knows anything about his personal history. But he is a golfer. And all those stories about his wagering on his game are true."

"I've talked to some fellows who say he is just a fair golfer —a 75 shooter."

"Listen, he plays a flexible game. If he's out with an 80 shooter, he'll shoot a 78. If his opponent gets down to a 70, he'll have a 68. I believe he can shoot down in the sixties about any time he chooses."

"What if he were tossed into a competitive field such as an open tournament?"

"Search me. But when the money chips are down he can play the game of golf. My brother plays with him every week. And Johnny Paolo is a low handicap man. But Montague can take him with all ease.

"One day I asked Johnny:

" 'What about him? Tell me the low-down.'

"And Johnny said, 'All I know is he is all right. And he can play golf. And he doesn't talk about himself.' "

All of which is just one more chapter in the many chapters about "Mysterious" Montague, who, some of the lads say, is the greatest golfer in the game. And who, some say, is just a fair golfer who got publicity by refusing to give any—that's the easiest way. Ask Miss Greta Garbo.

Anyhow, it was good seeing Pete de Paolo again. I saw him win at Indianapolis. And not long ago he was through with one of his books—"Wall Smacker." A new one is being written—the story of the mechanics.

Illustration 15-1 Article from the ***Atlanta Constitution****, April 22, 1937.*

The only news clip to be preserved is an undated one from the *Atlanta Constitution*, the essence of which is that De Paolo would not be driving, but managing a four-wheel drive entry. As for the great 500-mile race for which the De Paolo campaign had been mounted, once again there is no hint of coverage nor of pre- or post-race publicity in the file. What happened to the FWD Special was that Mauri Rose was not tempted away from Lou Moore, whose 4C270 he qualified eighth, only to go out on lap one hundred twenty-seven with a fractured oil line. For a driver, De Paolo came up with a rookie named Tony Willman, who qualified twenty-seventh in the FWD Special and threw a rod on lap ninety-five of the race. The winner, for the first of his three times, was Wilbur Shaw, in a creation of his own, powered by an Offy 255.

Ilustration15-2 De Paolo took over management of the Four Wheel Drive Auto Company car for 1937. To drive it, he chose a rookie named Tony Willman. It finished its Indy career on lap 95 with a broken connecting rod. (GBC ex FWDCO)

In the days immediately following the race, De Paolo devoted himself to the composition of a letter to Olen which purported to be his report on the race car. After strutting his stuff for more than two wasted pages he took it upon himself to draw up a blueprint for the Company's future, and his with it, all done with an air of gracious condescension.

June 7th,1937

Mr.W.A.Olen,Pres.
Four-Wheel-Drive Company
Clintonville,Wis.

Dear Mr.Olen:

This will serve as my report on the Four-Wheel-Drive racing car, which I managed during the last Indianapolis 500 Mile Race.

The uppermost thought in my mind at the time you offered the car to me, for the Indianapolis race, was the fact how impressed I was with the car similar to yours, which I drove in the Tripoli Sweepstakes, in Tripoli, North Africa in 1934,then later at the Avus Speedway in Berlin Germany, principally the safety factor of the car.

My first race with the car in Tripoli, after my retirement from active racing for a period of three years, my first thought--will the car handle well, and to my most pleasant enjoyment it did, to the extent that I qualified faster than my team mate, Lou Moore who had a much lighter car, with a faster engine, and too, my finishing ahead of Moore in the race. Needless to say the car created a great deal of interest while in Tripoli, by all the head officials of Mussolini's staff, principally the Governor, Italo Balbo, whom I gave a ride to in the car around the race course, picture which is shown in my book "Wall Smacker" he was greatly impressed with the Four-Wheel-Drive principal which I explained to him in his native tongue, and he went on elaborating on the Four-Wheel-Drive principal to all his associates there at the Governors palace, where I later dined with the entire group. It was quite obvious to me that he, Balbo, was very careful to assign different members of his staff to make notations on the Four Wheel Drive car, which his friend, DePaolo had brought over from America.

After the Tripoli event, I went on down to Berlin Germany, and there too, even moreso there was a great deal of interest around my car, particularly the morning of the race, when el Fuher "Hitler" and his staff gave a last minute inspection to the entire field of cars.

just prior to the start of the race, in a downpour of rain, when they asked me about my chances of victory in the race, I explained my chances were better than the other boys because of the fact that I had a Four-Wheel-Drive racing car, and this aroused a great deal of interest with all there, and needless to say there was a good bit of sketch making and notes on their "cuff" and what have you. I went on to explain to them the advantages I had over the other boys, because of my Four Wheel Drive, having power distributed to all wheels, and getting less slippage than they would get with only the rear wheel drive, particularly in the rain, they race there rain or shine.

After the race well on the way, I was proving the worth of the car, by passing most of the field, particularly on the turns, when I got up to 2nd place in the race, and directly in front of the grand stand where Hitler was supposed to be sitting, my motor threw out a couple of connecting rods, ruining my chances of victry, but no fault of the Four Wheel-Drive principal. A German officer later informed me that what ever flew out of my motor, just missed "Hiel Hitler" in his box, in which event know the Jews would have assigned a monument to the writer.

In handling your present car at the recent Indianapolis race, it was quite obvious to Jean Marcenac, my mechanic and myself, that we had our hands full in trying to get the car prepared for the race, in brief the car is not fit for racing in its present status, its not being balanced properly, not having sufficient power for the weight of the car, to cope with the speeds of the other cars of today.

This particular car was purely an experimental job on the part of Harry Miller, its original builder, at the cost of the Four-Wheel-Drive, Company, and looking at the entire set-up from this view, the car has truly performed most remarkably, as with all manufacturers when building new cars, there are a great many obstacles, or as we say in racing a lot of bugs to work out, until the car works out to their satisfaction, and it has been my experience with most of Millers cars, that it is necessary to practically redesign certain features before it is possible to win races with his cars.

Automobile racing is truly the experimental laboratory of the automotive industry, many mechanical features have been proven through the medium of racing, and in order for any manufacturer to gain recognition in the pleasure car field, or truck as is the case with Four Wheel Drive, they resort to racing, for in this manner they get front page publicity which money couldnt buy.

My idea of racing is to do it in a first class manner, to gain recognition with a car that will be a credit to the product which it represents, and not do it in a xxx hap-hazzard manner, therefore I suggest to the Four Wheel Drive Company, if they are interested in going ahead with a program on racing, in order that they will get further recognition, Nationally and Internationally, proving the Four Wheel Drive principal a great safety factor for pleasure cars as well as trucks, to set up a budget for a given amount, for example on a set up of two cars it would require $50,000.00 at the very least, which isnt a great deal when you consider the value of the great amount of National and International advertising you will garner. And with amount you will get the very latest in design and material in a team of cars that will be an asset to your organization.

After the cars have proven themselves in races both in America and over in Europe, under the supervision of the writer, we can outline a very extensive merchandising program, through all the various annals of advertising, and really put the story over 100% for The Four Wheel Drive Company, proving the safety of the product, and its introduction into the pleasure car, of which I am very much in favor of.

In the event all the members of the Four Wheel Drive, board of directors are in favor of such a plan, I will deem it a pleasure to consider entering your organization on a satisfactory basis to all parties concerned, to start with the company July 15th, to get started on the building of two racing cars of the latest design that will comply with specifications that will be International for the years of 1938,1939 and 1940,as this years cars will be obsolete for next years competition, making it a waste of money to do anything with the present car now owned by the company, or to get it fixed up for experimental purposes only..

I regret exceedingly my being unable to get there for your meeting on the 10th, when I would have liked to given you the set-up in a more personal manner, but I owe it to my present employers to comply with orders as given by them, until June 15th,when my contract will expire with them, and I will be ready to consider your proposition, as I have outlined in this report.

Assuring you that it has indeed been a pleasure for me to have represented your fine organization, and I trust it will be my good fortune to be with you in the future on a racing program first, and later the more serious side of business, merchandising, sales, and the ultimate objective of all business organizations, make money!!!!!!!

With kind regards to all there, I am

Most sincerely yours,

Peter De Paolo

Peter De Paolo

Hotel Detroit-Leland
Detroit.Mich. Until June 15th.

PdeP/zg

The telegram alluded to is not in the file. Apparently there had been discussion between Olen and De Paolo concerning rebuilding the existing car, rather than building one or more new ones.

June 30, 1937

Mr. W.A.Olen, President
Four Wheel Drive Auto Company
Clintonville, Wisconsin

Dear Mr. Olen:

My not getting an answer to my wire at an earlier date, I decided that possibly you were off on a trip somewhere and therefore would be unable to comply with my request on a meeting with you, as I mentioned in my wire, for this coming Friday. For this reason I made other arrangements to remain here in New York until after the race which will be run July 3rd, after which time I will drive through to Chicago via Detroit, where I will stop for a day or so.

Expect to arrive in Chicago the early part of the week of July 6th. If agreeable to you I would like to hold the meeting with Mr. Schmidt, your Sales Manager, the President of your advertising company and yourself on next Friday July 9th, when we can discuss the matters we both have in mind pertaining to the possibility of rebuilding your race car on the west coast and my holding meetings in behalf of your company, which I would be very happy to do.

In the event that the above arrangements will not comply with your activities, or any suggestions you might want to make, please write to me to the above address where I will be until July 4th.

With best wishes,
Sincerely yours,

Peter De Paolo

Another hallucinatory document. Between June 30th and July 14th a whole new plan was agreed upon between Peter De Paolo and the Four Wheel Drive Auto Company. The following legal instrument, signed by Olen, a lawyer, has the air of being a lawyer's nightmare. Item: It speaks of "the term of this agreement." But there is no term to it. One can only stand in amazement at De Paolo's powers of persuasion. The upcoming Vanderbilt Cup Race on Long Island,* plus overseas possibilities are foreseen.

WITNESS THIS AGREEMENT made this 14th day of July, 1937, between THE FOUR WHEEL DRIVE AUTO COMPANY, CLINTONVILLE, WISCONSIN, first party, and PETER DEPAOLO OF LOS ANGELES, CALIFORNIA, second party;

The first party agrees to loan to the second party the F.W.D. racing car, together with a two-ton (2) truck with platform body for the conveyance of the racing car from place to place as may be desired, and agrees to advance up to the sum of Five Thousand Dollars ($5000.00) for the purpose of rebuilding and remodeling the racing car.

The said PETER DEPAOLO is to supervise the rebuilding and remodeling of the racing car to make it suitable under the new laws and regulations relating to the Indianapolis Speedway and as near as may be possible to the international rules of racing cars so far as it applies to the Roosevelt race track.

Said PETER DEPAOLO is to supply an outline of the major changes that are to be made in the car for the approval of the Chief Engineer of THE FOUR WHEEL DRIVE AUTO COMPANY.

When the car has been rebuilt and remodeled, the said PETER DEPAOLO is to have the right to run it at such races as he may desire, and to receive all of the income from such races up to and including the race at the Indianapolis Speedway on May 30th, 1938 with the understanding that he is to pay the expense of entering the car in the race should the cost of remodeling and rebuilding the car exceed the sum of Five Thousand Dollars ($5000.00).

THE FOUR WHEEL DRIVE AUTO COMPANY agrees to pay such invoices as may be approved by the said PETER DEPAOLO for work or material in the remodeling and the rebuilding of the race car when they are submitted to the company or to advance a reasonable sum from time to time to be accounted for by receipted invoices for labor or material.

* Of which our Bill Burden was a driving force.

The said PETER DEPAOLO is; to have the right to take the race car and the truck to Los Angeles and keep it there during the term of this agreement.

Should the cost of rebuilding and remodeling the race car amount to less than Five Thousand Dollars ¢5000.00), then any remainder may be used by the said PETER DEPAOLO in entering the car or preparing it for such races, including the Indianapolis race.

THE FOUR WHEEL DRIVE AUTO COMPANY reserves the right, when the race car has been remodeled, to have tests run on such tracks or places in California as it may desire with the cooperation and under the supervision of the said PETER DEPAOLO.

Dated at Clintonville, Wisconsin, this 14th day of July , 1937.

Four Wheel Drive Auto Co.

Walter A Olen President

The birth of a race car: Here it is, in dramatic detail. De Paolo had come down off his podium and was in his element, mixing his talent with that of great men of the calibre of Goossen and Marcenac in the conception of a new super-race car. He had his teeth into what he loves, and the too familiar razzmatazz was gone.

August 9th,1937

Mr.W.A.Olen,President
Four Wheel Drive Auto Company
Clintonville,Wisconsin

Dear Mr.Olen:

Regret the delay in writing to you, giving full details as to what I have had in mind regarding the revamping of the present Four Wheel-Drive racing car, which has been due to my inability of getting to-gether with Mr. Leo Goosen, to discuss the possibilties of making the car lighter, faster, and more beautiful, as well as being safe..

It seems that everyone has started on some idea or another out here towards building new racing cars for next year, I have never witnessed so much activity in that direction, this early in the year, the reason for all this Mr.Olen is the fact that the new formula of rules and regulations will be good for the next three years, and the boys are wasting no time in getting ready, and they all seem to be quite lucky in having interested some "angel"(sponsor) to back them up financially, so that you will see a wonderful selection of racing cars at the line up next year, most of them being improved models over the present type of cars, really no radical changes.

Mr.Goosen, Jean Marcenac and myself have had several meetings together in the past week, to thoroughly talk over the problem we have with your racing car such as it is. Under separate cover I am mailing to you a sketch on what I have in mind, as follows:

1--A one seat car, with somewhat European design, blunt front end, sort of tear-drop effect for better stream-lining.

2--Widening of the tread, from 56 inches to 62 inches, for safety, the car having less tendency to "wander", which all drivers dread, like fog to an aviator..

3--Submitting the power to all four wheels, from both ends of the motor, using the present transmission for the rear-end drive, and redesigning the front end with an over-running clutch, giving the effect as a front-drive, thus still maintaining the four wheel drive principal, and the elimination of a lot of weight with the present transfer case set-up, which is better for trucking purposes and not racing cars, because of the additional power it requires on account of friction through the many gears.

4--New four wheel hydraulic brakes, similar to the European design, which will enable the car to make a good showing at the Roosevelt Raceway, as well as European racing.

5--A new 4-cylinder motor to comply with the present formula, rather than rebuilding the present motor which is now in the car, as it almost amounts to building a complete new motor, all with the exception of a new crank-case. It would be my suggestion to install a new crank-shaft in the motor and sell it, and apply the amount we get towards a new motor, as I am sure we can get approximately $1500 for the motor, and the new one will cost $2,850,making the net cost on the new motor only $1350.

Here are the new features and their approximate costs:

New Body and chasis:- consisting of:

A-newly designed body
B-new frame side rails
C-all new cross members
D-New 6 gall.,oil tank
E-New Gas tank-35 gallons
F-New lines gas & oil throughout
G-supports for new center drive steering
H-Entirely new instrument board
I-all underpans to conform with new body.

Total cost submitted by Clyde Adams who builds most all of the Miller cars - - - - - - - - - - - - - - - - - - 1500.00

New 4-cylinder motor by Offenhauser:-with improved features over the present 255 cu. in. motor now in chasis:

A-Larger main bearings
B-More connecting rod bearing surface
C-Larger wrist pins
D-Longer pistons-eliminating rocking effect.
E-Newly designed, stronger crank-shaft, larger counter weights, better balanced and hollow drilled.
F-New style crank case, better ribbed for strength.
G-cylinder block with longer stroke.

This motor complete with bell housing, rear engine support plate, fly wheel, intake manifold, two Winfield carburetors, and exhaust leader flange:

Total cost complete submitted by Fred Offenhauser - - - - - -$2850.00

Change over and redisgning of the four-wheel-drive as shown on blue-print, drawn by Leo Goosen who has designed most all of Harry Millers cars:

Drawings-patterns-castings etc., as quoted by Mr.Goosen,
Approximately- -$1750.00

Four Wheel Hydraulic brakes, with dual master cylinders, and the maximum size of drums etc., to give most braking surface, as quoted by Bendix Company Los Angeles:

Approximately- -$600.00

Complete summary:-
Body & chasis- - - - - - - $1,500.00
Motor- - - - - - - - - - - - - - - - -2850.00
New front end- - - - - - - - 1,750.00
Four -wheel-brakes- - - - - - 600.00
Total cost- - - - - - - - - - - -$6,700.00

This cost in addition to mechanic's salary of Marcenac and myself,for over a period of 5 months time at the rate of $400.00 per month for the two of us, as I will work right with Jean as well as supervise the entire job, the total labor cost I would place at $2,000.00

$6700.00
2000.00
$8700.00

In the event you should desire to have me fix up the present motor, it will cost approximately $1350.00 which would just about off set the price we could sell it for, unless we decide to keep it as a spare motor, which isnt a bad idea, then this amount should be added to the above total:

$8700.00
1350.00
$10050.00

Approximately $10,000 for the entire deal, would cover everything, with the possibility of some left over, through the careful supervision of the writer on the entire job, how-ever even should the undertaking require the full amount as quoted, I consider this quite a reasonable amount, as we would have a new racing car, with a spare motor, and equipment we would be proud to compete with, either in this country, abroad, or anywhere racing might exist.

In presenting this deal to you Mr.Olen, I have quoted all prices at the maximum, with the exception of the motor deal of $2,850.00 which is the actual cost as quoted by Offenhauser, and the new body and chasis deal as given by Clyde Adams, all other cost are more or less of a rough guess, therefore it might not cost this much, even though I would much rather quote a higher amount and play safe, than say a lower amount then have to resort for more finance.

I ask for no money, only the salary to Marceanac and myself, otherwise all bills will be forwarded to your office for your okay, or any other manner which you might suggest, the only current money I might ask for would be for minor details, such as getting set-up in some shop to do our work and transportation money, as fuel for car etc., or as your corporation would say, petty cash, all of which I would submit an accounting to you for, as you would like it, monthly or weekly.

I am enclosing your agreements which I am returning unsigned, so that you may revise them in accordance with this letter, and if you feel it would be better for me to fly back there, and give us an opportunity to discuss the entire matter in full, I would be very happy to do so, and if all goes well, I will personally drive the truck back with the racing equipment, so that Jean and myself can get started to work.

When thinking this matter over Mr.Olen, please bear in mind that all these other fellows who are building cars at this time have appropriated budgets ranging from $12,000.00 to $15,000.00 for their cars, and too your records there will show what the original cost was on the car you now own, I believe $15,000.00 when it was an experimental job, and that was some three or four years ago, too old for a racing car, and now we can get an entirely new one for approximately $10,000.00

I am anxiously awaiting your reply so that I may arrange my affairs accordingly, and I am looking forward to the possibility of flying back there to see you personally and close the whole deal up in a satisfactory manner to all parties concerned.

With best personal regards, I am

Sincerely yours

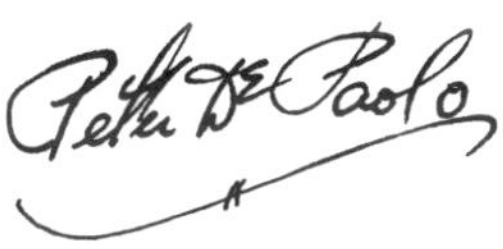

Peter De Paolo
<u>New address</u>
3308 Tica Drive
Los Angeles,Calif.

The new concept, as drawn by Goossen:

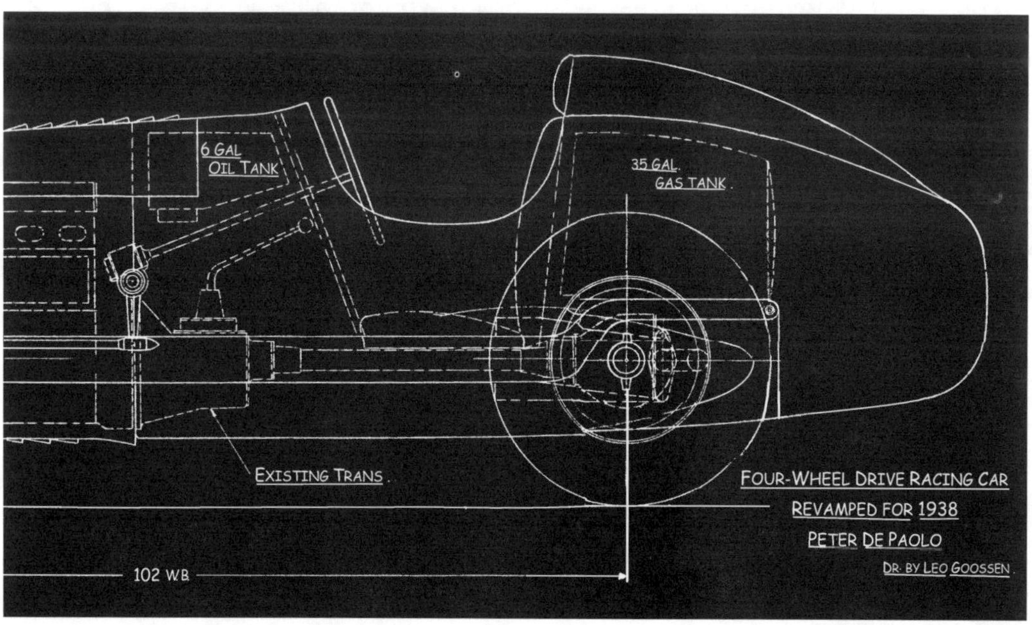

Illustration 15-3

Four months passed in silence, according to the file. Olen's communication to De Paolo on December 9th is missing, as are all of his other communications. Then comes De Paolo's last love letter to Olen, leaving nothing unstated.

Dec.12th, 1937

Mr.W.A.Olen,Pres.
Four Wheel Drive Auto Company
Clintonville,Wis.

Dear Mr.Olen:

This will acknowledge your kind letter of Dec.9th, and the check enclosed for which I am very grateful.

Its a genuine pleasure to get your letters Mr.Olen, for as I read through them I enjoy the feeling of getting a message from one whom I consider almost like a Dad:

I follow a famous Father
And never a day goes by,
But I feel that he looks down on me
To carry his standard high!

I have placed you with a group of men in my mind, who are in a category of those who are doing all within their power, towards the good, and safety of mankind, far above the average:

Fred S. Duesenberg
Charles M. Schwab
Harvey S. Firestone
W.A.Olen

Today I am leaving for Indianapolis where I will pick up a new LaSalle car, and drive to Los Angeles and join my family whom I haven't been with for the past two months.

If you feel that I can be of any assistance to you or your company while I am on the Pacific coast, please advise me to the address given below, I expect to be on the west coast until Feb.1st,after which time I will start on a new series of meetings for the Hastings Mfg. Company.

The thought I have in mind in regards to your company, the possibility of your advertising, and sales promotion department organizing a series of meetings for the west coast this winter, and possibly more of them after the Indianapolis race, in June or July, for the west coast lets say 4 meetings, Los Angeles, San Francisco, Seattle, and one in Arizona, perhaps Phoenix.

Have your district managers in the above mentioned cities arrange all necessary preparations for the meetings, inviting fleet owners, and purchasing agents to attend, all potential buyers etc., feature Peter De Paolo as the guest speaker, with newspaper publicity, and radio talks in advance of meetings by the writer. Have other members of the Four Wheel Drive Company present at each meeting, touching on engineering, and some of the outstanding units which have been produced by the company,telling about the different parts of the world where your product is sold, and a general resume on the romance, and progress made by your company, which is all very interesting, yet more or less kept a secret. I can pep up the meeting in my talk to them as the last speaker, telling them about my racing experiences, the factor of safety which I enjoyed while driving the Four Wheel Drive car, the Germany race incident etc., and at the conclusion of the meeting serve a buffet lunch, and give the Four Wheel Drive group an opportunity to visit with the buyers, making appointments with them etc...

If you are interested in a series of these meetings, I will gladly sandwich my time with the Hastings Company in your behalf, my agreement with them is that I will not represent any other PARTS manufacturers for the year of 1938, making me eligible for any deal I might make with an automobile or truck manufacturer.

I am enclosing a letter which I recieved from Mr, Johnson, president and general manager of the Hastings Company, which is self explanatory, also a letter from the president of Borg-Warner Company, commenting on one of my many talks, thought perhaps you might need these for evidence in selling the idea of any expenditures for meetings of this kind to your board of directors, which all comes under the heading of better advertising, the finest medium of garnering new business, as Wrigley would say "tell them, and tell them often", money well spent.

My fee for such meetings will be $250 each, with all travelling expenses paid by your company, Which I am sure you will agree with me is very reasonable, taking into consideration the cost of magazine and trade journal advertising.

If the first series of meetings prove successful, as I know they will, we could go ahead with more of them after the Indianapolis race next summer, incidentally this year at Indianapolis I will be a member of the writing staff for the Indianapolis News, conducting a column to the activities at the speedway each day two weeks prior to race day, "Speedway News" by Peter De Paolo, needless to say I will occassionally give mention to my old love The Four Wheel Drive racing car, the arrival in town of Mr. Olen, or any other members of the company etc., you betcha!!!

Meetings after the Indianapolis race could be set for the following cities, Washington,D.C., New York City, Boston, and Portland Maine? ...

I am enclosing a newspaper mat which can be used in connection with these meetings, if you care for any more of them please advise me, enclosed you will also find a section of the front page from a morning Chicago Tribune, with a picture attached, saying "meet Pete DePaolo at the Hastings booth", a copy of the Tribune with this sticker attached was placed under every door of rooms in the following hotels, Stevens, Palmer House, Drake, Sherman House and the Knickerbocker, a bit of good merchandising I would say ... the show was a big success, many new jobbers were signed up, and too,I autographed approximately 1000 copies of my book "Wall Smacker" which were given out with the compliments of the Hastings Company.

I would like very much to get your re-actions to my suggestions for meetings in behalf of your company, as quickly as possible, so that I may advise the

Hastings Company enabling them to arrange accordingly for the start of my next series of meetings with them.

With best personal wishes, I am
Cordially yours

Peter DePaolo
3308 Tica Drive
Los Angeles.Calif

COPY

HASTINGS MANUFACTURING COMPANY
Hastings, Michigan
INTER-COMPANY CORRESPONDENCE

To	Peter DePaolo	From	A.E. Johnson
	c/o Detroit Leland Hotel,		
City	Detroit, Michigan	Date	June 15, 1937

Dear Peter:

Sorry we did not have an opportunity to get over to Flint, and while we haven't yet heard bout the meeting, we feel very sure that you had a good crowd at Flint. Do not expect very good crowds at Bay City or Saginaw but they may surprise us.

Now, Peter, we desire to avail ourselves of the option on your services for a period of four months in 1938 at a salary of $1250 a month.

We feel that your work has been very much worth while, and it is our thought also, that with the experience we have had this year in arranging meetings we will be able to do a better job the coming season.

All of the jobbers we have heard from where meetings have been held have been very well pleased. We also feel that the publicity we receive from using you in connection with our advertising and our literature has been effective.

We will be pleased to have you stop in Hastings and see us on your way west if you find it convenient.

Very truly yours,

Hastings Manufacturing Company

A. E. Johnson,
General Manager.

AEJohnson:B

COPY

BORG-WARNER CORPORATION
310 South Michigan Ave.,
Chicago, Ill.
June 5, 1937

Mr. Peter DePaolo
c/o Mr. Fred Willis,
Chicago Motor Club,
Chicago, Illinois.

Dear Pete:

I have just returned to Chicago from Detroit, following our dinner at Indianapolis last Saturday night, to the success of which you contributed so much.

I want to thank you very deeply for your kindness in talking to the boys at the dinner, and want to repeat that you put up about the swellest after-dinner talk I ever listened to.

If you are going to be in Chicago Monday, I would be very happy if you could find time to drop in and see us at our offices on the 27th floor of the Straus Building, 310 South Michigan Avenue.

I would like again to have an opportunity to thank you, and to talk over some other matters with you which might be of interest.

With best regards, I am

Yours sincerely,

Chas. S. Davis,
President. [Borg-Warner Corporation]

CSD/ef

And Olen's reply, in which he was as borne away as he had been by Barney Oldfield, five memory-filled years before.

December 16, 1937.

Mr. Peter DePaolo
3308 Tica Drive,
Los Angeles, Calif.

My dear Mr. DePaolo:

If my letter of December 9th put any enthusiasm in you, your letter of December 12th just put a lot of new life, faith and confidence in myself. I am sure that I will have no other Christmas present that will do me so much good as your letter.

I am returning the letters from Borg-Warner Company and the Hastings Mfg. Company,

The question of using you on lectures will be given serious consideration but it will probably take a little time to work this out.

Wishing you a merry Christmas and a Happy New Year, I am

Yours very truly,
THE FOUR WHEEL DRIVE AUTO CO.

W.A.Olen President
ELV
Encl.

On this exact note ended the Four Wheel Drive Auto Company's adventure in the very far-out world of American oval-track racing on the National Championship level. There were other adventures to come.

Chapter 16

The Rebirth of the Miller 45° V-8

Art Pillsbury, the draconian boss of the A.A.A. Contest Board in the western United States, had busted Bunny Phillips around 1930 for driving in "outlaw" events. His period of probation ended as the 1936 season approached, and he acted accordingly. He had rebuilt the engine of his blown 138 cu. in./2261 cc Type 35B Bugatti, but had yet to attack the major task of rehabilitating the scattered 308/5048 unblown Miller which he yearned to use in its place. He left Dick Loynes's employ early in 1936 specifically in order to hit the A.A.A. Championship trail. Before leaving, and with an eye on the future, Bunny purchased from Dick all of the wooden patterns for the V-8, all of the many spares that he had for that engine, and, as we know, the entire hoard of Miller drawings. Those for the V-8 were the only ones that he really wanted, and he eventually sold the rest to Mark Dees.

Bunny Phillips's Bugatti-propelled racing career was filled with adventure, good and bad, through late 1939, when he engaged the services of Leo Goossen to render his V-8 fit to face the coming season. Master machinist and mechanic Phillips worked closely with Miller's master engineer. With the intention to eliminate the lower-end problems of the past, they redesigned the crankshaft thrust-bearing arrangement. Originally, the thrust had been borne by number three of the five main bearings. They deleted that feature and designed a much beefier rear main to take over that burden. Then, the overwhelming problem of rod-bearing failure had been due to heat generated by the side-by-side rods rubbing against each other, their steel turning blue from the heat and the babbit, of course, melting. The solution they chose was to reduce very slightly the width of the rod big ends in order to provide space for a separator web, machined integrally on the journals of the necessarily new crankshaft. Having inspected the engines of the two four-wheel drive cars after the 1932 Indianapolis race, Harry Miller, on June 2nd, sent a long report to Leo Goossen, which is reproduced in its entirety by Dees. Under modifications to be made, one finds, on page 368A of Dees's second edition, "Engine to have rods separated on shafts so they will not rub each other." Bunny said that he knew of no other V-type engine in which this technique had been used, although Goossen did use it again in 1938 when designing a new crankshaft for the ex-Burden Miller 303 V-16. Both Miller and Goossen lacked experience with V-type engines, be it with fork-and-blade or side-by-side rods. Tens of millions of the latter have been built without this having been a serious problem. In Bunny's case, the modifications rendered Miller's 308 V-8 trouble-free. Why?

Beginning in 1938, the Indianapolis Motor Speedway adopted the A.I.A.C.R. International formula, which limited the displacement of unblown engines to 4.5 liters/274.59 cubes and of blown to 3.0/183.0. Therefore, during the redesign of the Phillips mule engine, the 4.0-in. stroke was reduced to 3.5, matching the bore and giving a new displacement of 269.4 cu. in./4315.8 cc.

Between this engine and the final-drive housing of his Type 35B Bugatti, Phillips installed a Ford truck four-speed gearbox and torque tube. He designed new body panels with modern lines; they blended into the car's original, pointed tail section, which he retained. Bunny trailered this hybrid all over the vast country, his tow car being a 1935 Hispano Suiza convertible sedan.* He quarried ahead of the works Maseratis at Indianapolis in 1940, but was bumped in an exercise of track politics. Furious, he stormed to Detroit, where a dirt-track 100 miler was taking place that same weekend. He finished a fine second there, campaigned all over the East, and finished thirteenth in the A.A.A. Championship. In 1941, he came in thirteenth at Indianapolis and scored a rather remarkable eighth in the Championship. He then joined the war effort as a manufacturer of aircraft components. Following that detour, he became the legendary Bugatti specialist of the United States. He carefully conserved most of his old treasures and, in the winter of 1994–95, turned his old Bugatti-Miller over to a more youthful friend, Dr. Peter Williamson of New Hampshire.

The original engine of the Four Wheel Drive Auto Company car has been for many years the property of Bill Goodwyn, who has a museum in Frankfort, Indiana.

*At this writing the Hispano Suiza is owned by Jacques Harquindeguy of California, who drives it regularly.

Chapter 17

Historic Coincidence: Bugatti Four-Wheel Drive

That Harry Miller and Ettore Bugatti both should produce four-wheel-drive cars—and of almost the same displacement at practically the same time—is a phenomenon that must be interpreted with great prudence. Neither man was spying on the other and, although Harry Miller's influence on Ettore Bugatti is an open book, it goes back to front-wheel drive, from which Ettore Bugatti either failed or did not deign to grasp the reality of what was in his hands.

Ettore Bugatti's son Roland assured me that, when his father acquired the two Packard Cable Specials in September of 1929, he was driven by two motives. One was to study the Miller engine, which was capable of astonishing specific power output. The other was to study the structure and behavior of the Miller system of front-wheel drive. This, Roland said, was not because Ettore Bugatti wanted to build front-wheel drive cars, but because he recognized here the key to four-wheel drive, which really interested him. And the four-wheel-drive race cars that he built were not seriously intended to be that. They were rolling laboratories for learning how to build high-speed, all-terrain military vehicles. As with Harry Miller, two were built, and probably in the same sequence of mule and production prototype.

Although Ettore Bugatti copied the Miller 91 engine faithfully from the base of its paired cast-iron cylinder blocks upward, in the case of front-wheel drive he royally rejected every detail of Miller's way of solving problems. Considering Bugatti's offenses in order of magnitude, the most staggering was his failure to insist upon CV U-joints to which he had enjoyed a privileged introduction, thanks to his possessing the Millers. René Dreyfus set a new record for the La Turbie Hill Climb at the wheel of a Type 53 Bugatti four-wheel drive in 1934. He told me that those cars were the most difficult to drive that he ever had known. He did not drive up the mountain near Nice, he said. Curves had to be taken in a series of arm- and shoulder-breaking straight-line spurts. This was because of the absence of CV U-joints.

The design of the Type 53's front end was in harsh contrast with traditional Bugatti practice whereby, as with Miller, beauty of form was inseparable from function. This design was the handiwork of Antonio Pichetto, who had come to Molsheim from Turin. There, he had worked in the consulting office of famous engineer Giulio Cesare Cappa, formerly of Itala and, before that, of Fiat. Knowing that there is a Cappa-designed front-wheel drive racing Itala monoposto of circa 1927 in the Biscaretti Museum in Turin, one had but to glance at its front end, or photos of same, to see where the Pichetto approach had come from. To this was added a central change-speed gearbox, with transfer gearing and differential included in the same housing. From it, drive shafts extended to offset front and rear final-drive/differential housings.

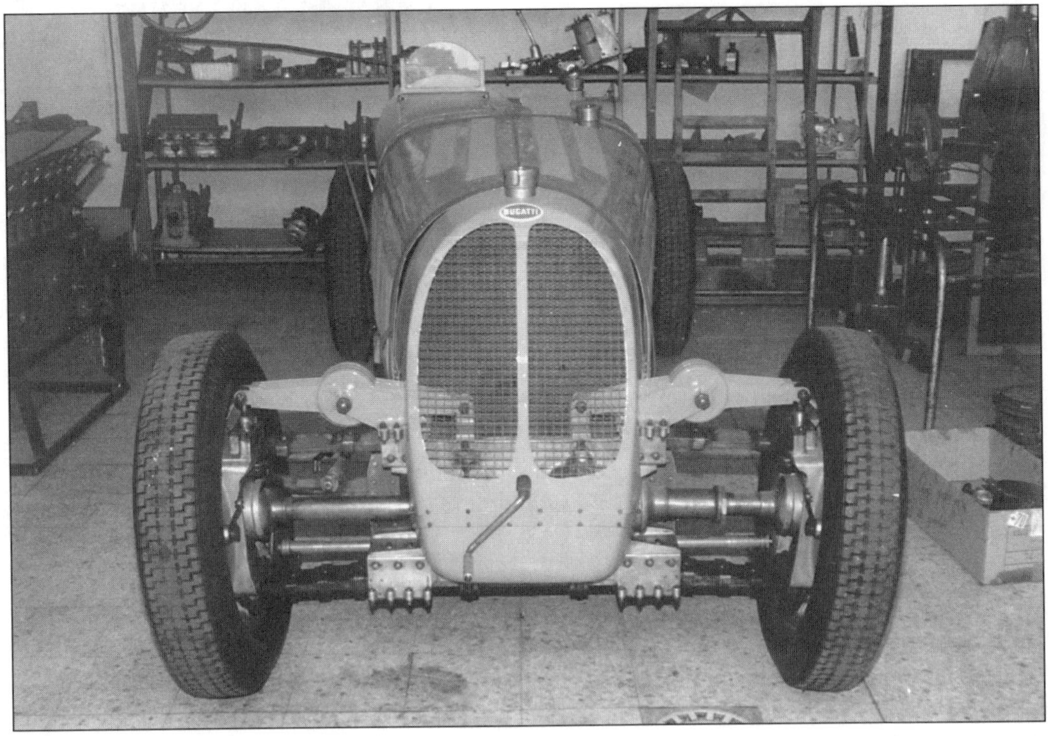

Illustration 17-1 The front suspension of Ettore Bugatti's four-wheel drive Type 53, by upper and lower transverse leaf springs, is unlike that of any car he built before or after. The lack of characteristic finesse also is striking. (GBC)

Illustration 17-2 According to René Dreyfus, who drove the Type 53 brilliantly, its outer front universal joints are not of the constant-velocity type. Presumably, a single cardan joint does the job, after a fashion. (GBC)

Illustration 17-3 The Type 53 uses a conventional live rear axle, with offset center section mounted on typically Bugatti reversed quarterelliptic springs. (GBC)

Illustration 17-4 1925 Itala four-wheel-drive monoposto. Transverse leaf springs are covered by protective sheaths. Bugatti T53 front end was of essentially identical design. (Carlo Otto Brambilla)

The National/Schlumpf Museum in Mulhouse contains the frame of one of these cars, along with its correct front-wheel-drive assembly and steering gearbox, as well as what seems to be its correct rear-axle assembly. Its wheels are Type 51, rather than the radically different Type 53. The other car, the property of Uwe Hucke and restored by the Laurent Rondoni organization, is virtually like new and participates frequently and impressively in vintage car events.

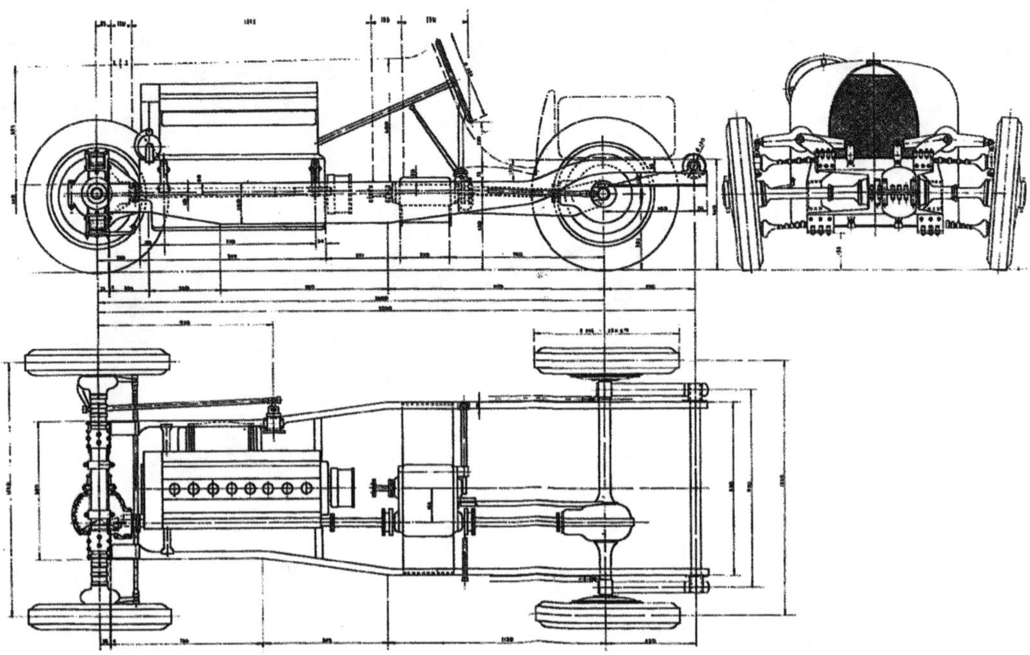

Illustration 17-5 This Super Car Graphic drawing of the Bugatti by Yoshihiro shows clearly the similarities and differences between the Bugatti and Miller four-wheel-drive cars that were built more or less simultaneously, 6000 miles apart. (GBC)

Chapter 18

Bill Milliken—Science and Adventure with Vintage Iron

When Walter Olen wished Peter De Paolo a Happy New Year at the end of 1937 he probably did not anticipate the grave recession that was about to begin and would hit the Company hard and extend well into 1939. The FWD Special gathered dust, and then came the boom years of World War II, when all-wheel-drive trucks and jeeps became an outstandingly decisive form of armament. The race car slept on, to be wakened in 1948 by a thirty-seven-year-old Massachusetts Institute of Technology (MIT) graduate.

William F. Milliken was born in Old Town, Maine, in 1911. As a kid he was attracted by exciting cars, but his greatest interest was in aviation. He was nineteen and majoring in mechanical engineering and physics at the University of Maine when he designed, built, and flew his own airplane. "I managed," he recalls, "to build into it every known form of aircraft instability and, having landed it on its back, I was impressed with the problems of stability and control. Later, at MIT, I was fortunate in taking courses from Professor Otto Koppen, the real pioneer of aircraft stability and control analysis in this country."

Milliken received his bachelor's degree in math and aeronautical engineering from MIT in 1934. He spent most of the war years with Boeing as assistant head of flight-testing. Then, after a stint with Avion (Northrop), in 1944 he moved to Curtiss-Wright, Flight Research Department in Buffalo. After the war, in 1946, that facility was transformed into the Cornell Aeronautical Laboratory (CAL), known today as Calspan. Bill remained there until 1976, serving successively as head of flight research, then director of the full-scale division, which was composed of flight research and vehicle dynamics. Finally, he became director of the lab's transportation division and its executive engineer. Bill himself tells how cars entered his professional career:

"In 1945 I happened to see an MG TB in New York City and talked its owner into selling it on the spot. This led me to the Sports Car Club of America (SCCA), which at that time had about thirty members. I soon swapped the MG for a Type 35A Bugatti, which had been raced prewar in Automobile Racing Club of America (ARCA) events. I took it to Pike's Peak in 1947, because there were no other places where a sports car driver could race at that time."

Bill finished sixth at the Peak that year, against some very hardened experts. He was pretty intrepid himself. For a long time the little Grand Prix Bugatti was his sole transportation and, in it, he drove home to Maine, did a three-hundred mile trip to Albany in midwinter, and drove to sports car races as they started being held, including the very first events at Watkins Glen and at Bridgehampton. The car lived in Cornell's flight hangar at the Buffalo

Airport, where it was maintained and fussed over by the lab's highly qualified technicians, who thoroughly shared Bill's enthusiasm for good high-performance machinery. Certain of them chose to take their vacations at the time of the Pike's Peak climb, in order to act as the car's pit crew. Bill recalls, "We were engaged at that time in some really pioneering work in aircraft dynamics for the Air Force. We were, in fact, the co-inventors of the variable-stability aircraft. As an engineer working in aircraft dynamics, it was pretty easy to become interested in the handling characteristics of automobiles under competition conditions.* A couple of us were so curious about car handling that we went to Detroit and visited several car companies to find out what they knew about it. The result was that we obtained a substantial contract from General Motors to transfer aircraft technology to the automobile. We did this first in the Flight Research department and later started a Vehicle Dynamics department and, finally, a Transportation Research division. I gradually worked out of aircraft control and into automobile control, although I take advantage of aircraft technology to this day."

To Milliken's eternal credit, his experience with the Bugatti at Pike's Peak in August 1947 led him to draw an original conclusion, which he proceeded to pursue to the end of its implications. The gravel-strewn course mounts 4960 feet in its 12.42-mile length and contains a hundred and ten curves. As he warped his way through them, descending as well as climbing, he saw that the key problem was traction and that the key to improved traction was all-wheel drive. He began searching for a car thus endowed, and through SCCA contacts he learned of the forgotten four-wheel-drive Miller that lay in dead storage in Clintonville. Using his Cornell credentials, he wrote directly to Walter Olen, who suggested to Bill that he come to the factory and explain what he had in mind. Bill told me:

"I met with him and his son Don, and explained that my interest was a combination of racing and engineering. Walter Olen agreed to the loan of the machine for Pike's Peak if I would pay all the expenses (they amounted to about $3500) and report in engineering terms on our experience. That was our bargain. They trucked the Miller to Buffalo, where we kept it in our flight hangar, and began the preparations, which included changing the front and rear axle ratios from 3.38 to 4.30 (which the car has to this day).

"Working with the people at the Four Wheel Drive Auto Company was a fine experience. The situation there was reminiscent of Bugatti's in Molsheim. Here was a small automobile company, started by a single strong individual and dominated by his family. It was far and away the largest industrial establishment in the area, and its employees were of the craftsman sort. There really was no production line and each vehicle was to some extent unique. They had an excellent model shop, in which racing work could be carried out under serene conditions. I got to know top engineers Don Olen and Jack Fabian intimately, and Jack's residence became my home in Clintonville."

This accord, based initially upon a single fact-finding expedition to Pike's Peak, was so mutually rewarding that it endured for five years. In addition to extensive testing carried out

*Another contributing factor may have been Bill's accident proneness. In his fifteen years of driving in about a hundred and fifteen races, it was only with the Miller that he did not have a serious accident. Milliken's Corner at Watkins Glen received its name during the historic first race there, in 1948, when Bill landed his Type 35 on its back in that bend.

Illustration 18-1 Unloading the Miller Four Wheel Drive at its new home away from Clintonville, the Cornell hangar at Buffalo. A test circuit was improvised around hangars and on runways. (RMRC ex WFMC)

Illustration 18-2 Members of the Cornell Flight Research group at Buffalo Airport, 1948. Bill Milliken is second from right, Bill Wilcox in center, and the Type 35A Bugatti is the one with which he scraped sixth best time at Pike's Peak the year before. The Miller still carries its 1937 Indy numbers. (RMRC ex WFMC)

Illustration 18-3 Pike's Peak, 1948. Cornell Flight Research mechanic Bill Wilcox, Bill Milliken at wheel, talking with the Four Wheel Drive Company's Herman Larson. Note the big, heavy-duty lever-type friction front shock absorbers. (RMRC ex WFMC)

on the runways of Buffalo Airport, it included the following competitive activities, with Milliken always at the wheel:

EVENT	DATE	RESULT
Pike's Peak Hill Climb, Colorado	Sept. 1, 1948	Transfer case cracked just above Glen Cove (halfway up).
Watkins Glen Grand Prix, New York	Sept. 17,1949	Magneto failure. Repaired, finished far back
Mt. Equinox Hill Climb, Vermont	May 21, 1950	Fastest time of day and course record.
Edenvale Airport, Ontario, Canada	May 25, 1952	First in first event, second in second, fastest lap.
Mt. Equinox Hill Climb, Vermont	June 14 & 15, 1952	Second, broke record three times during this weekend
Burke Mountain Hill Climb, Vermont	July 4 & 5, 1952	Exhaust valve broke, 1/4 way up.
Glenwood Hill Climb, New York	May 23, 1953	Fourth. Very rough course.
Niagara Falls Speed Trials, New York	Sept. 20, 1953	Event closed prematurely.

The object of this prolonged investment of time and other valuable consideration was to assess the four-wheel-drive principle under the conditions imposed by road racing. All parties concerned were fully aware that the car had been conceived primarily for oval-track use. However, it happened to be the only all-wheel-drive racing car available, eliminating the question of alternative choice. Among the problems that Milliken and his chief mechanic Bill Wilcox had to deal with were:

STEERING RATIO. The Miller was equipped with a Ross cam-and-lever steering box with a drop arm that measured 6.60 in. between centers. The ratio that this provided was perfectly satisfactory for the Speedway, and Brisko did not complain about it on mile dirt tracks. On any sinuous road course, however, it would not do at all. Milliken wrote to the Ross Gear & Tool Co. in Lafayette, Indiana, and was able to obtain an arm with a length of 8.50 in., thought to be the maximum available. It speeded up the steering considerably, but for the Peak it still was no match for the nimbleness of the Bugatti. Ross billed Milliken $2.33 for the precision-machined forging, which presumably included shipping. Before the Peak, Bill managed to find a 9.75-inch arm but even with it the steering was still too slow.

STEERING LOCK AND RADIUS. In addition to the slowness of the steering, due to the limited angular deflection that the Miller's front outer U-joints were capable of, the steering lock was and remains very limited. Its turning radius was about twenty-five percent greater than that of the Type 35 Bugatti. This was a source of grave concern before the Miller ever left Buffalo, but, at the Peak, Milliken developed a technique that permitted him to squeak through even the tightest turns. He remembers vividly today that "going up, where the turns end toward the hill and under power, was not too hairy. But coming down, when the turns end toward the 'outside,' the brink, the only way to get around them was to have the car in first gear and to hit the throttle hard enough to break the tail loose. That was really pretty scary."

Illustration 18-4 Milliken cornering on gravel at speed. Note the perfectly normal, balanced attitude of the Four Wheel Drive car, in spite of right front wheel being well off the ground. (RMRC ex WFMC)

Illustration 18-5 A very powerful photo. Speeding through a series of S-bends, the car continues to track normally. In order to achieve this result on the rutted surface of dirt and gravel, it was necessary to lock out the central differential. (RMRC ex WFMC)

Illustration 18-6 Contrast the poise of the four-wheel drive with the cornering attitude of the 1948 Pike's Peak winner. Grotesque is the word. With his physicist's knowledge of vehicle dynamics, Milliken knew that four-wheel drive was the answer when he first went up the peak in a rear-drive car. (RMRC ex WFMC)

TRANSMISSION. As everyone knows, in oval-track racing this contraption serves to get into top gear well up on the torque curve, after which the job is to stand on it and turn left. Only a very elementary sort of gearbox is required, and the Miller had one with three pairs of spur-cut gears, spaced widely and unevenly. What, of course, is wanted for road and mountain racing is a boxful of cogs that, in the interest of smooth, fast, and non-destructive gear changing, have their respective ratios chosen and spaced with care. Purism aside, synchro-mesh is a blessing, for man and machine. Milliken eventually became a pioneer of torque converters in road racing, but with the Miller he was condemned to live with the crash box. Taking the train from the East Coast to Denver, he practiced, mentally, double-clutch down- and up-shifting all the way. Ace mechanic Wilcox, the final recipient of the results of wear and tear, credited Milliken with having become a competent exponent of double-clutch artistry. The ratios of the car's original box were 1.00, 1.22, and 2.15 to one; the spread between second and third being awful for road racing use. Jack Fabian eventually equipped the box with a set of 1.43 gears in place of the 1.22s, making the best of a trying situation.

TRANSFER CASE. Although spectacularly out of its design element at Pike's Peak, the Miller performed extremely well in practice and until about half-distance in its timed sprint up the mountain. At that point, its ever-problematical transfer case came apart at its seams. Milliken's promised report* to the Four Wheel Drive Auto Company—which turned out to be monumental—stated, "Unlike Indianapolis, we were accelerating and decelerating most of the time, and during the acceleration periods we were attempting to operate at maximum engine torque. On deceleration we made use of the engine, in part, as a brake. It seems likely that the fatigue loads on this assembly [the transfer case] are magnified by operating in the [indirect] gears. It is also possible that appreciable shock loads were sustained by this assembly during the process of shifting up and down, although it was felt that the double-clutching technique gave reasonably smooth shifts."

The original Miller transfer case had been a disaster. Now this one, an in-house Four Wheel Drive Auto Company design, had begun to deceive and would go on doing so. As disappointment followed rebuild, and with participation at Watkins Glen scheduled for 1949, a very interesting development took place: a new search for old four-wheel-drive Miller parts. The Four Wheel Drive Auto Company's Jack Fabian addressed this letter to Leo Goossen on July 9, 1949. Among other things, he expressed an interest in "the original transfer case," obviously being unaware of its horrendous history.

*The FWD at Pike's Peak—Confidential report on the preparation and participation of the Four Wheel Drive Auto Company's "FWD Special" in the 1948 Annual Pike's Peak Hill Climb with an examination of the technical problems encountered in this competition, by William P. Milliken, Jr., January 1, 1949, Buffalo, N.Y.

July 9, 1949

<u>AIR MAIL</u>

Meyer and Drake Engineering Corp.,
2001 West Sage Ave.
Los Angeles, 44, California

Attention - Mr. Leo Goosen, Chief Engineer

Dear Mr. Goosen:

We are preparing our "Four Wheel Drive Special" for road racing. The car, built by Miller in 1932, was originally powered by a V-8, but this was subsequently replaced by the present four cylinder 255-cubic inch engine. Last September Mr. W. F. Milliken, Jr. drove this car in the Pike's Peak Hill Climb during which, unfortunately, the transfer case failed. We understand that he contacted you about the car at that time and also once later when he discussed the possibility of bringing the engine up to a competitive position with the more recent models.

Although we are definitely interested in this modernization of the engine, there is at present a somewhat more pressing problem. Mr. Milliken is hoping to drive the car in the Watkins Glen (N.Y.) road race on September 17th, and we feel that the present clutch and transmission are not satisfactory for this type of usage, even if the transfer case were in perfect order. It was thought that perhaps you might know where an evenly spaced four speed synchronized transmission and heavy duty clutch, or an epicyclic gear box, could be obtained which would take the punishment of sustained operation in the lower gears and frequent shifting such as encountered at Pike's Peak, and yet be adaptable to our machine with a minimum of labor. Perhaps others have attempted this same conversion, or there is a commercially available unit which somewhere near fits this description. Perhaps you build one.

We realize, of course, that the time element is such that in order to get the car to Mr. Milliken in time for sufficient practicing and familiarization before the race we may well have to be satisfied with much less than our ideal for this season. By next year we hope to have the car completely rebuilt. So anything we do now, if it entails any major expense, would have to be a permanent change.

It might be more to the point to just change the ratio of second gear in our present three speed box for this season. The present ratios, 1.00:1, 1.22:1 and 2.14:1, do not give very good spacing for road racing. A second gear of about 1.44:1 would be far preferable. Unfortunately we have no design data on this transmission so it is difficult to conjecture as to its life expectancy in this sort of service. Perhaps youcould shed some light on this question, as I understand you were with Mr. Miller at the time the car was designed. We would also be interested to know if by any chance the original transfer case might still be available.

If you know of any such clutch and transmission which would fit between our engine and transfer case and which we could get in about three weeks time I would very much appreciate your telephoning me collect about it.

Very truly yours,
THE FOUR WHEEL DRIVE AUTO CO.,

G. J. Fabian
Senior Design Engineer

MMM

Three days later, Goossen replied to Fabian, referring him to Dick Loynes concerning the patterns for the primordial case.

July 12, 1949

The Four Wheel Drive Auto Co.
Clintonville, Wisconsin

Attention G. J. Fabian
Senior Design Engineer

Dear Sir:

Replying to your letter of July 9th. We regret being unable to give you much help in respect to a heavy duty clutch and transmission for the four-wheeler, especially in the time available. The writer layed-out the original clutch and transmission for Miller, but it being some time ago my memory is rather uncertain about the strength of the parts for your requirements, and we have no drawings. It might be worth a chance, however I would hesitate to commit myself either way. It would of course be possible to change the gear ratios by making new gears and perhaps at the same time make them heavier.

Regarding a stock transmission and clutch it looks as though you could find something among the many commercial transmission and clutch manufacturers in the east and would suggest contacting them.

In reference to the original transfer-case pattern, Mr. Richard Loynes, Ocean Blvd, Long Beach, Calif. had all the four-wheel drive patterns at one time and it may be they are still available.

With best wishes for your project and please write if we can be of further service.

Yours truly

MEYER & DRAKE ENGR. CORP.

Leo Goossen, Chief Engineer
LG:es

On July 16th, Fabian informed Milliken that Goossen had given "us a lead as to where we might track down the original FWD Special patterns, so that if anything else gives way we won't be completely lost." It was here that he suggested the band-aid treatment for the original gearbox. And it took him until August 9th to pursue the Loynes lead with a letter. It was summer vacation time, after all.

July 16, 1949

Mr. Wm. F. Milliken, Jr.
Manager, Flight Research
Cornell Aeronautical Laboratory
4455 Genesee Street
Buffalo 21, New York

Dear Bill:

I know you are disappointed about the transmission situation, because I am myself. However, I cannot feel it would be prudent to spend the time and money to adapt the Maserati transmission to our car. In the first place the car that it was designed for, as nearly as I can determine, weighs about half what ours does, and was only a two wheel drive to boot. That indicates that we might impose something of the order of four times the load on it as did the Maserati. From the shaft sizes which you gave me, the shaft stresses in the Maserati box would be approximately two and one half times as large as for the same engine torque, assuming geometrical similarity. And when you add to this the fact that the engine torque of the Maserati 4C-16 valve engine is only 144 foot pounds against our present 215 foot pounds and expected 262 foot pounds. When the engine is rebored this winter, I can't say as I'd feel safe about trying to use it. I hope I'm not being a damper, but after analyzing the situation it seems like hopping out of the frying pan into the fire.

I got a very nice, but unfortunately nugatory from a transmission standpoint, letter from Goosen of Meyer & Drake. He did give us a lead as to where we might track down the patterns for the original F.W.D. Special, so that if anything else gives way we won't be completely lost.

How about my putting a new 2nd gear ratio in the present box, if nothing else shows up, which will give a ratio of about 1.43:1? This would give at the rear wheels 15 MPH/1000 engine R.P.M. which rather evenly splits the 21.0 in high and 10 in low. Or would you rather have some other ratio? The time is now getting so short that we'd better look at all the possibilities.

I think that this winter, in the redesign, I can dream up a combination transmission transfer case that will give us anything we want. There's plenty of room for it. We'll talk it over later, but keep it in mind. We can get exactly what we want for next year this way.

I've been going over the Mercedes report and note that their spring rate was 224 #/in both front and rear, so I will attempt to achieve this in the F.W.D. Special by August 20th.

The material has arrived for the new clutch plates and the patterns for the transfer case are being brought up to date. Arnie is on vacation now so he'll be in good shape when tho parto otart rolling in.

Sorry about the transmission, but I hope you understand. The foreward prop shaft, which cuts into the present bell housing doesn't simplify things either.

I'll return your books in a few weeks if you don't mind their being away that long. I'm finding them very interesting, especially the German report.

I'll try to keep you posted on our progress.

Very Sincerely,
THE FOUR WHEEL DRIVE AUTO CO.,

G. J. Fabian
Senior Design Engineer

GJF:ct

It was not until the end of August, with Watkins Glen just seventeen days away, that Loynes replied to Fabian. The delay may have been due, at least in part, to working out an accord with Bunny Phillips, who had bought the engine patterns and drawings from Loynes over a decade before. Loynes apparently had scrapped the rest of the four-wheel-drive patterns, along with other cumbersome treasures, "when I acquired all the Miller patterns and drawings."

August 31, 1949

AIR MAIL

The Four Wheel Drive Auto Co.
Clintonville, Wisconsin

Attention: Mr. G. J. Fabian

Dear Mr. Fabian:

I have your letter of August 9 regarding the patterns for the four wheel drive race car. I have been out of town, so I am a little late in answering.

I do have all the patterns and drawings for the V-8 cylinder motor that was used in this car. They are all in very good condition. I do not have any of the patterns for the various chassis parts, for the reason that when I acquired all the Miller patterns and drawings we did not have sufficient room for storage and every thing was reduced down to the patterns for the various engines at that time.

There evidently is another one of these cars in Pasadena as I provided the owner with a cylinder block some months back.

If you are interested in these engine patterns and drawings, I would be very happy to provide you with a list of everything I have.

Yours very truly

R. R. Loynes,

R. R. Loynes

CENTER DIFFERENTIAL. At the Peak, the Miller handled badly on the low-coefficient surface until it occurred to Bill Wilcox to try locking the center differential. Our old acquaintance Herman Larson, who had observed most of the efforts of the company car at Indianapolis, was able to report that doing so had not worked there. Doing it at the Peak transformed the car. Fully aware of the delicacy of this point, Milliken expressed himself thus in his formal report to the factory:

"An unfortunate feature of the transfer case failure is that it could be associated, and in fact might even have been precipitated in part by the fact that the car was operating with locked center differential at the time. In our experience at Pike's Peak, locking the center differential had a profoundly beneficial effect on the handling qualities of the machine, and it is to be hoped that this factual conclusion will be viewed entirely on its own merits, as distinct from any structural implications of the case failure. The writer well appreciates the importance of the center differential in the application of the four-wheel drive principle to machines operating on average road surfaces. As explained in the text [of this report], surface conditions at Pike's Peak, at our speeds of operation, were far from normal and locking the center differential proved to be the most effective way of actually maintaining the drive on all four wheels, with all that that implies in handling and acceleration."

Milliken also warned that "Before the center differential was locked, any misjudgment in application of power usually resulted in either front or rear wheels losing traction and spinning. If this occurred on a turn, the car would usually go into an uncontrollable spin or excessive plow. During the time trials, with unlocked center differential, the car inadvertently spun twice to a dead stop, through between 100 and 180 degrees. During the actual race, with locked center differential, the same course was traversed at a substantially higher speed, with no signs of instability. Mr. Larson had warned that the high-speed behavior with locked center differential at Indianapolis was unsatisfactory, the car tending to spin, and generally being 'tricky' to control."

Traction was rather optimal at the Speedway. At the Peak, it was like driving on shot, marbles, BBs.

"With the differential unlocked," Bill continued, "it was easy, when braking down for a corner, to lock the rear wheels and hence break traction and slide. The improvement with locked center differential was most noticeable when for any reason braking was required on a corner." Bill arrived at the interesting conclusion that, with the center differential unlocked, the car was overpowered. Locked, "the acceleration and deceleration characteristics were improved to the point where the car became underpowered."

SUSPENSION. Milliken told me, "The suspension of the Miller is extremely stiff. We checked the natural frequency, undamped, and found it to be about 212 cycles per minute. Compare this with about 100 to 120 cpm for most racing cars and 50 to 70 for passenger cars. Because of this very high spring stiffness, which is fine for smooth tracks, the Miller proved to be a handful on a really rough surface. There, at speed, it is not difficult to get into a PIO [pilot-induced oscillation], with the vehicle yawing back and forth as you put in out-of-phase steering, hoping to ride out the decreasing speed and amplitude before you lose it."

On this subject, springmaster Marvin Edwards, the force behind once-famous Hollywood Spring & Axle Co., at 6009 Sunset Blvd., told me long ago about his clientele. It included savvy race-car owners who, on buying a new Miller, would bring it directly to Edwards, to be fitted with springs that actually flexed. This speed secret never reached Clintonville.

STORAGE BATTERY. It had become standard practice in early 1930s oval-track racing to jump-start engines by means of a battery that was located handily in a car's pit or on its pit wall. The Miller four-wheel drive used a 24-volt battery to spin its on-board Bosch starter. The high probability of spinouts and stalls at the Peak dictated keeping the battery aboard the car. It rode in the space reserved for the riding mechanic.

Milliken was convinced from the outset that the four-wheel-drive Miller, properly sorted, could acquit itself outstandingly in both road and mountain racing, up to and including victory. When victory did come, in the SCCA's Mount Equinox Hill Climb in 1950—the car's third outing in Milliken's hands—there was joy in Clintonville, as well as among the CAL crew. There began to be serious thought of a production sports cars, built by the Four Wheel Drive Auto Company and, of course, featuring four-wheel drive. At 6.25 miles, Equinox was about half as long as the Peak. The climb was just 65 percent of the Big One (Pike's Peak), but it was steeper, with an average gradient of 9.75 percent against the Peak's 7.58. It was and is a hell of a test of both car and driver. The only threats in the Large Sports Car Class were the XK-120s of Briggs Cunningham and Bill Spear. Spear got up the mountain in 7 minutes and 53.0 seconds, while Briggs was just 0.3 seconds slower. There were only two cars in the Unlimited Class: the four-wheel drive and George Weaver's three-liter blown Maserati monoposto. Far more maneuverable, and with a greatly superior weight-to-power ratio (7.7 lbs/hp versus the old Miller's 11.7), it should have been able to run circles around the heavier car. But only its rear wheels were powered. So, when Milliken set the record by charging up the course in 6 minutes and 59.4 seconds, against Weaver's 7 minutes and 08.0 seconds, the advantages of four-wheel drive were there for all to see.

That was not all. On Milliken's first practice run that day, the Miller's gearbox began popping out of 2nd gear, meaning that at least half of each ascent had to be made with one hand on the steering wheel and the other holding the shift lever in place. Then, on his final record-setting run, when about a mile from the summit and approaching the final series of switchbacks that led to it, Milliken downshifted to 2nd. As he let out the clutch there was a chattering scream as the teeth peeled from their respective gears. Bill declutched instantly, upshifting to top gear, only to discover that it, too, had died. It must have taken some deft and desperate double-clutching and throttle synchronization to get the bottom cogs to engage. Thus he carried on to the top, enjoying the luxury of being able to keep both hands on the steering wheel. During the famous downshift, from the corner of one eye he saw the tach needle hit 6800. That is really winding an engine with a 4.50-inch stroke; in fact it equals about five thousand feet per minute piston speed. The car's entry for upcoming Bridgehampton was withdrawn, on the grounds that it had been "somewhat overextended" at Equinox, thus setting a record for understatement.

As a matter of fact, the car was out of circulation for two years. After a well-deserved rebuild, Milliken drove it in a couple of races on the Edenvale Airport circuit in Ontario, Canada, on May 25, 1952. Three weeks later saw him back at his Mount Equinox stamping ground. There, he and the Miller were second fastest in the two-day event, during which they broke the course record three times. The final three outings of this pair were uninspiring, all the desired lessons had been learned, and Milliken and the FWD Company turned their attention to an entirely new four-wheel drive project (A.J. Butterworth 4WD), a special constructed in England.

As we have seen, Milliken continued with CAL and Calspan until 1976, becoming increasingly interested in the application of aircraft technology to the control and stability of all types of earthbound motor vehicles. At this point, he founded his own consulting firm, Milliken Research Associates, Inc., in Williamsville, N.Y., taking with him such tried and true talent as Dr. Hugo Radt and David Segal. The company specializes in analyses and testing in the areas of automobile stability and control, handling, aerodynamics, and tire mechanics for a large customer base of car and tire manufacturers and well-known racing teams. Bill's right-hand man in this activity is his son, Douglas, forty years old at this writing. Together they wrote a book, which was published recently by the Society of Automotive Engineers (SAE). *Race Car Vehicle Dynamics* is an encyclopedic treatment of the subject, the fruit of Milliken's several decades of specialization within it. Looking back today on the handling of the original Miller four-wheel drive with which he shared so many adventures, he says:

"The Miller was, of course, a superior car under low road friction coefficients. Movies taken during practice at Watkins Glen shows the car going around a corner in the rain as if on rails. Then comes Briggs Cunningham through the same corner in a Inter Ferrari. He spins out and winds up with the front wheels in the ditch.

"As you know, the Miller is a very symmetrical car. It has identical De Dion suspension on each end, and the center of gravity of the car is just slightly aft of the mid-point of the chassis. Since the lateral tire forces act slightly aft of the center of the tire print, the resultant lateral tire force is almost directly in line with the centrifugal force from the center of gravity. Assuming equal tire sizes and pressures the car is essentially neutral steer by design. It feels very much that way, and tends to corner with a feeling of being on rails. It certainly doesn't tend to hang its tail out, like high-powered rear drives. We developed a very safe and interesting cornering technique at the Peak, by coming close in to the corner and then applying the power and drifting out as we progressed around the corner. The application of power to the four wheels tended to drift the car sideways. The cornering technique certainly was very comfortable for an amateur, as compared to the dirt-track slides of the rear-drive cars.

"The Miller has one pronounced handling problem: trailing throttle oversteer. Thus, if you are in a turn and take your foot off the throttle, the car tends to "duck in," or tighten up the turn. One can, of course, compensate for this with the steering, but in the heat of a race that may be awkward. In those days we didn't thoroughly understand this problem and we made many experiments to try to alleviate it. We moved the center of gravity forward and aft with lead weights, without much improvement. And we also played with tire pressures. Trailing throttle oversteer is a 'trim change.' It now is well understood, and suspensions are designed to control it. Basically, it is due to the fact that when one lifts his throttle foot in a turn, there is a transfer of weight forward. That increases the tire lateral forces on the front and decreases them on the rear which, in turn, tends to yaw the vehicle into the turn. Every car has it to some extent. The Miller has a fairly high center of gravity, and that tends to exaggerate the effect."

Read Bill and Doug Milliken's eminently readable book and become an expert yourself. In addition to being particularly sensitive to trailing throttle oversteer, the Miller four-wheel drive is the same with regard to that old tabu, braking-in-a-turn. One can perfect one's understanding of these phenomena by reading the following references in the Milliken book:

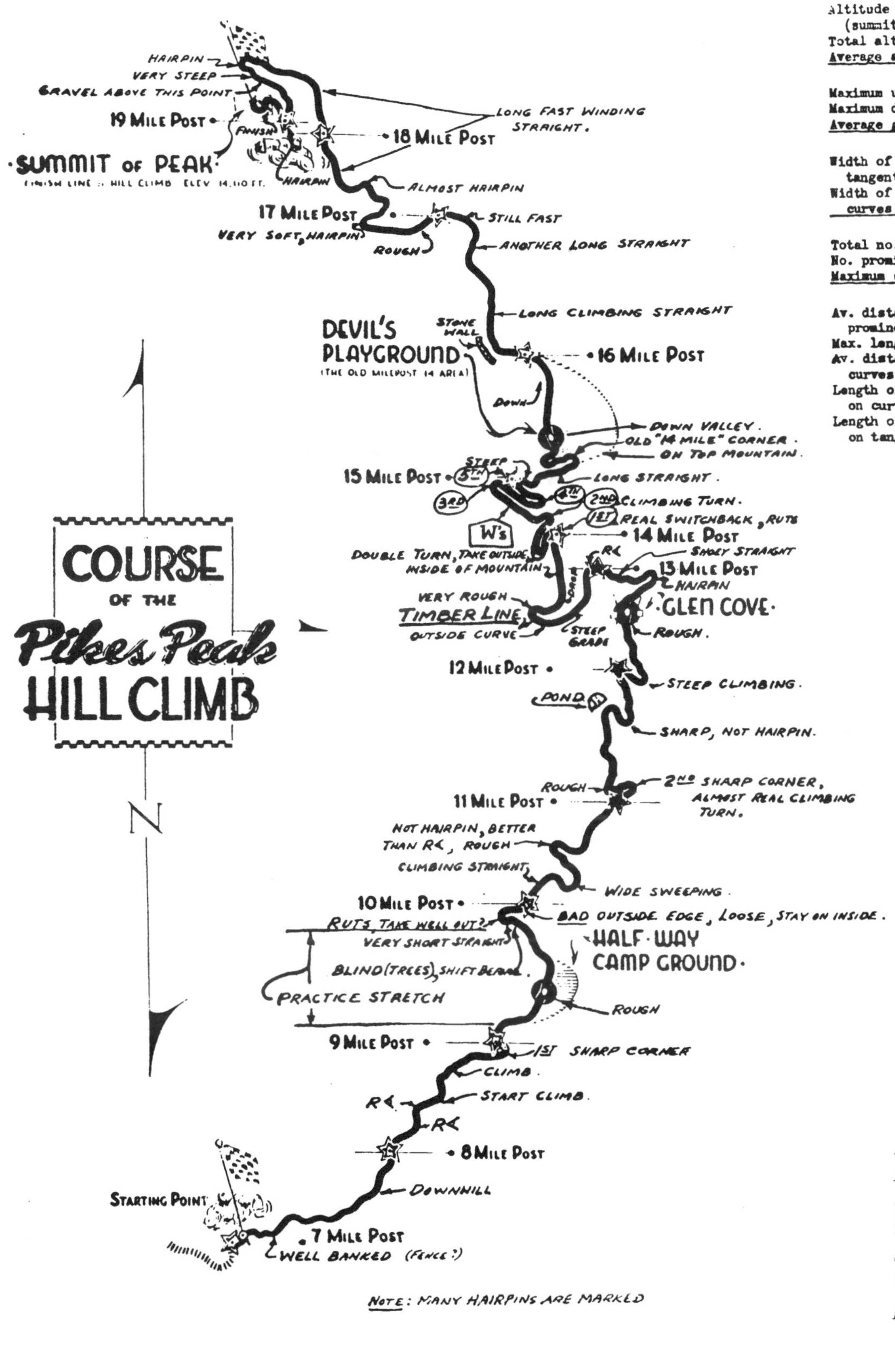

COURSE STATISTICS

Total distance	12.42 mi.	
Altitudes		
Altitude of start	9150'	above s.l.
Altitude of finish (summit of peak)	14,110'	above s.l.
Total altitude change	4960'	
Average altitude	11,630'	
Grades		
Maximum up grade	10.5% (6°)	
Maximum down grade	5% (approx. 3°)	
Average grade (up)	7% (4°)	
Roadbed Widths		
Width of roadbed on tangents	20'	
Width of roadbed on curves	20'-50'	
Curves		
Total no. curves	110 (approx.)	
No. prominent curves	65 (approx.)	
Maximum curve	146°	
Distances		
Av. distance between prominent curves	1000' (approx.)	
Max. length of tangent	1082'	
Av. distance between curves	600' (approx.)	
Length of course on curves	42% (approx.)	
Length of course on tangents	58% (approx.)	

Illustration 18-7
Milliken's sketch of the course with personal notations, from his comprehensive report to the Four Wheel Drive Auto Company. (RMRC ex WMFC)

1. Discussion on page 354 beginning with, "An interesting feature of Figures 9.5 and 9.6 is ... "

2. Pages 397 and 398 in the chapter on Chassis Set-Up: paragraphs 24 through 29.

Milliken is the author or co-author of many technical papers that have been presented to and published by SAE and IME, the British Institution of Mechanical Engineers. A work of his which is a milestone in the literature of four-wheel drive is his "The FWD at Pike's Peak,"* the confidential report which was a major part of the bargain that he made initially with Walter Olen. This tome consists of nearly two hundred bound, mimeographed, single-spaced typewritten pages of analytical documentation. Of the few made originally, one copy; no doubt the Company's; is preserved in the E. Dean Butler archives. It has been of the greatest value to this work.

Note by Bill Milliken

As Griff Borgeson points out, initially FWD (Walter Olen) had little experience in racing and this led to misunderstandings with Harry Miller of both a technical and financial nature. When I approached FWD much later, there were no such problems. Thus, initially there was no financial exchange as the car was loaned to me and I bore the expense of campaigning it. Furthermore, FWD and my interests coincided in that we were both enthusiastic about furthering our knowledge of 4WD handling and performance.

I found Walter Olen and his son, Don Olen, very direct and easy to work with, as were the FWD engineering personnel. Herman Larson went with us to Pike's Peak and Jack Fabian was with us at Watkins Glen in 1949. Jack and I formed a close friendship, and when he left FWD some years later, we hired him as a member of Cornell Laboratory's Vehicle Dynamics Department, where he was responsible for the tire tests on the six-component, on-road tire tester. He stayed with Cornell Laboratory until his retirement.

Jack was still at FWD when we became interested in the Butterworth 4WD machine in England. The FWD Company acquired it and we developed and raced it for a number of years under FWD's sponsorship. Always known as "The Butterball," it was developed by Bill Close into an exciting machine.

Walter Olen was a remarkable man. When the company he started pre-WWI, expanded to meet the demands of that war, he was under great pressure as the top administrator. To maintain his health, he would spend his spare time in the woods chopping down trees and other strenuous exercise. When I met him he was in his eighties, still effectively directing his company and pushing the 4WD principle.

*The FWD at Pike's Peak Confidential report on the preparation and participation of the Four Wheel Drive Auto Company's "FWD Special" in the 1948 Annual Pike's Peak Hill Climb with an examination of the technical problems encountered in this competition, by William P. Milliken, Jr., January 1, 1949, Buffalo, N.Y.

Chapter 19

Changing Caretakers

I began to gather data relating specifically to the four-wheel drive Millers in mid-1965 and wrote to the Company for that purpose. Publication publicity manager Arthur Danley replied on July 19th with the expected potted history of their car. He said that their photo negatives had been lost, but that they had prints of the usual static views of car, driver, and riding mechanic, taken at the Indianapolis Speedway each year. I had expressed an interest in technical shots, if available. In his reply, Danley said, "regarding that suspension, it appears that you race car authorities have a language all your own. What is de Diou?"

The next input was a press release dated September 20, 1965 stating that "Seagrave, major producer of commercial and custom-built fire fighting equipment, is wholly owned by FWD Corporation." This was indeed news to me. The point of the release was that the Timmons Metal Products Co., of Columbus, Ohio, would be building fire apparatus for the Seagrave division of the Four Wheel Drive Auto Company. The next and final communication from Danley was dated October 4th, and accompanied the long-awaited Speedway static shots. I had asked about the possibility of getting photos of the outer U-joint interiors. He said, "The race car is actually the property of the Four Wheel Drive Foundation, of which Mr. Olen is president. He didn't take too kindly to our disassembling the race car for this purpose." According to the company history, *The Four Wheel Drive Story* (McGraw-Hill, N.Y., 1954), "the Foundation was created in 1948 for benevolent purposes."

What I had been able to gather from all sources was threadbare. I tried to reactivate the project in 1980, which I outlined in a letter to Danley or his successor. Response: nil. After Bob Rubin instigated this book project I wrote to the company's public relations manager on March 28, 1994, advising him of the fact and requesting a few elementary bits of reference material. No reply. After two more futile attempts, I wrote to the Clintonville Chamber of Commerce on August 2nd, asking if the Company still existed and, if so, what was its present address. A reply by return mail from Linda Marotz of the Chamber, said that the Company was still very much there, and gave me the name of executive secretary Wanda Tellock. I sent her this fax message:

Campagne Mirail, 13 August 1994

Ms Wanda Tellock
FWD CORPORATION
105 E. 12th Street
Clintonville, WI 54929 USA

Dear Ms Tellock:

Herewith copy of a letter which I addressed to your PR director last March. Having received neither a reply nor my returned letter I recently wrote to the Clintonville Chamber of Commerce in order to verify wheter your firm still exists and, if so, where. Linda Marotz of the C of C kindly referred me to you. A small amount of cooperation from your organization would add greatly to the quality of what promises to be a very good book and one which will be of distinct benefit to your company. My requests are:

1 and 2 of my letter of 28 March remain unchanged.

3 - Production from the beginning to the end of WWII will take care of my needs.

4 - RE FWD MUSEUM BOOKLET, hereinafter called booklet, and counting the front cover as page 1, can you supply me with a good copy of the original newspaper clip? This one has been reset in modern type, which spoils the intended effect.

5 - A good print of the photo on page 5.

6 - Good print of photo on page 8.

7 - How many FWD vehicles were shipped to Europe during WWI? Were they all Type B?

8 - When did the Seagrave merger occur and who absorged whom?

9 - The economic environment is very significant in this historical work, for which reason I would like to know the company's financial balance for each year, from founding to end of WWII.

10- Fact sheet or equivalent describing the company as it is today.

11- Two copies of the Booklet to be mailed to me, plus one each to

Mr. Robert M. Rubin, Director, AIG TRADING GROUP, One Greenwich Plaza, Greenwich, CT 06830. Mr. Rubin is a distinguished collector and the present owner of the FWD Miller Special racing car, to which pages 1 and 2 of your Booklet are devoted. He is the instigator of my Smithsonian book on Miller, as well as of the work now in progress.

Monsieur Paul Berliet, President, Fondation de l'Automobile Marius Berliet, 39 avenue Esquirol, 69003 Lyon, France. Mr. Berliet is a world authority on the history of the automotive industrial vehicle, of which the Foundation's museum contains a very great number of examples. He is a sincere admirer of your firm's products and probably would be very happy to add a Type B to the Foundation's collection if one could be found. His restoration facilities are excellent, as are those of Mr. Rubin.

Much time has been lost, and I really hope for response to the above at your earliest convenience Thank you, and cordial regards

(signed) Griffith Borgeson

She replied seventeen days later:

August 30, 1994

Griffith Borgeson
Mirail
8424 LaMotte D'Aigues
France

Dear Mr. Borgeson:

I have received your letter of 8/13/94. It will take some time to find answers to all of your questions.

Some of the material has been destroyed before my arrival on the scene which is about 7 years ago. Seems there was a cleaning done in several of the buildings here and a lot of old information was destroyed. I apologize for this hasty action, but I will send on what I can to fulfill your requests.

I am aware of the sale of the FWD Race car to Mr. Rubin. In fact, we are questioned quite often as to the disposition of it at the present time. Has Mr. Rubin been able to assemble the race car?

The last we had heard it had been taken apart before the sale to Mr. Rubin and we often wonder about it.

I will be sending as much Information as I can for you within the next week along with some suggestions as to where additional information could be obtained.

Thank you for your inquiry.

Very truly yours,

FWD CORPORATION/
SEAGRAVE FIRE APPARATUS, INC.
(signed) Wanda D. Tellock
Executive Secretary
WDT/

Bob Rubin brought Ms. Tellock up to date. I sent her courteous reminders on September 30th and October 19th, but we never heard from the Four Wheel Drive Auto Company/ Seagrave again.

The FWDCO Museum

This company institution, which is open to the public, continues in full swing in Clintonville. At this writing its current twelve-page descriptive booklet devotes its first two interior pages to "the Miller FWD Special." Despite this reinstatement of Harry Miller's name, the photo chosen to illustrate the historic machine has all of its lettering obliterated. Below the photo and opposite a page full of racing performance summaries is the statement, "All of the Vintage FWD's described in this Booklet are on permanent display at the FWD Museum, Clintonville, Wisconsin." The style resembles that of ad manager W.N. Hanson in the 1930s.

Robert J. Boudeman

Buck Boudeman, then of Hickory Corners and now of Richland, Michigan, electrified the old-car world when he first unveiled his impeccable recreation of Miller's immortal 1917 Golden Submarine. That this event took place at Hershey in 1983 is well documented, but it is one of the few events in Boudeman's remarkable career that is easy to nail down. He is a great historian. He lives history very close to its sources, thanks to the dreamboat-load of wonderful vintage racing cars and related treasures that he has had the taste and determination to seek out and possess over the years. He planes over such trivia as just when a given event, such as an acquisition, happened. Of course, it was recorded by others that he was born on November 12, 1941. When he first became aware of the existence of the four-wheel drive Miller in the Four Wheel Drive Auto Company Museum nobody knows. He says that it was about ten years before he got the car. So when did he get the car? He doesn't know, and so far no scrap of paper has turned up to settle the point. Mark Dees says that it was in 1984, but that doesn't jibe with other inputs. Let's assume that it was in 1970, when Buck was going on twenty-nine. He's a great big guy, with a dry, country-boy sense of humor, and a drawl to go with it.

"When I learned that the car was there," he told me, "I started making inquiries at the Four Wheel Drive Auto Company plant. I called all the way up the chain of command until I finally got the president of the place. He said, 'Son, the only way you're ever going to get that car is by buying this company.' This went on for several years and my wife finally said, 'Why don't you just take no for an answer? Those people are never going to sell you that car.' Miller-collector friends Chuck Davis and Dave Uihlein laughed at me and said the same thing.

"But I kept on. About once a year I'd call up and go through the whole thing again. Finally, I called one more time and got the president. It was a different guy. He said, 'You know, we've merged with Seagrave, and we'd like to have some historic fire trucks in our museum, instead of that old race car. We'd consider selling it.'

"So, I made him an offer, and got close to closing the deal. But they talked with some of the long-time employees, who didn't want to part with the car. They considered it to be part of their history, and management decided to go along with them. I was very disappointed, went home from Clintonville, and mulled it over. I decided to make them one more offer and I said to them, 'Rather than just take the money, you could put it into an account and use it for financing an annual old-timers' reunion.' That's how I got the car.

"Dave Uihlein had invited me to come and view his collection if I ever got over Milwaukee way. So on the way back from Clintonville, with the FWD car on a trailer, I made a detour to his place and he showed me all his beautiful cars; the big Miller, the Sparks-Thorne Big Six, the Indy Mercer and all the others. In the course of the tour he said, 'What's under the tarp on that trailer?' I said, 'Oh, it's an old race car.' When we were all done and I was ready to leave, Dave said, 'Let me see what you've got on the trailer.' He had his pipe in his mouth. I pulled back the tarp and he just about bit the stem off his pipe. After a long silence he regained his composure and said, 'How the hell did you ever get that out of the Four Wheel Drive Auto Company?'

"Back home," Buck continued, "I pulled the crankcase sideplates off and looked inside; there was no babbit lying on the screens. I pulled the cam covers off and the cups weren't grooved. The oil tank is part of the front-drive unit; I took the radiator off and washed out the oil tank, flushed it out and made sure there were no chips or anything in there. I dumped some new oil in it, put a fresh magneto on the engine, and it fired up like it had run the day before. I drove that car around the block and I was stunned. If you were blindfolded and got in that car and drove it, you would not know it was four-wheel drive. It drove like a baby buggy. It is the most forgiving car that I have ever driven. I took it down to Winchester, Indiana, and drove it on the high-banked half-mile track there. Emil Andres showed up and probably put fifty laps on it there, taking everybody for a ride. The car just ran flawlessly.

"When I bought that car, one of the conditions was that I would bring it back to the Four Wheel Drive Auto Company plant for, I think, their 75th anniversary.* I did, parading it through town and driving it around their test track. While there, the idea came to me to drive over to Menomonie and see if I could find Harry Miller's birthplace. So I loaded the race car on the trailer and drove over there, unannounced. I pulled into the town and inquired about the birthplace. They told me where it was and I drove out there and looked. While I was there I met a gal from the local Historical Society. She said, 'You know, we've been trying to open the Menomonie Historical Museum for two and a half years. Tomorrow is the big day. Would you mind parking the race car in front of the museum?' Which of course I did. There never had been a Miller race car in Menomonie before, but this one showed up just as the museum opened. When you stop and think about the numerical possibilities ... it was like it was meant to be.

"It is a lovely car, and I really love it. It's sort of a window into Harry Miller's mind. It's all right there. But then I eventually got in a position where I had to sell something, I sold the four-wheel drive to Bob Rubin."

Griff: "When was that?"

Buck: "Gosh, I don't know. Why don't you ask Bob?"

* That would seem to have been 1983, which does not jibe with other data.

It will come as no surprise to the reader to learn that the front-drive unit, the transfer case, the radiator shell, and other pieces from the Miller four-wheel-drive No. 1 mule found their way into Bunny Phillips's trove of four-wheel-drive material. Nor to learn that the elements just cited were purchased from Bunny by Buck a few years after he had acquired the car from the Four Wheel Drive Auto Company. He still has them. He is the man who recreated the Golden Submarine around an original engine. With these original parts, what might he create next?

Historic Places

As for the museum of the Dunn County Historical Society in Menomonie, its present curator, John M. Russell, says that perhaps Buck was misinformed concerning the grand opening. The event must have been one of those that, since 1970, has marked the annual opening and the spring thaw. This cadence should change soon, since the Society has been the recipient of a five-acre tract on which to build a new museum that, better in every way, will permit year-round operation. Construction awaits termination of a big drive for funding, and the contributions of Miller enthusiasts are particularly welcome. An important Miller section, of course, is planned, and there has already been talk of inviting the home-state Miller back to Menomonie for an opening, which is sure to be grand. It is to be noted that, while this book has been growing a page at a time, Harry Miller's birthplace and childhood home has been nicely restored and now forms part of the Society's circuit of historic sites in the region.

During the same period, the U.S. National Park Service has decreed that Harry Miller's old 'Malibu Ranch' is "eligible for listing in the National Register of Historic Places for its association with Harry Miller." The mountainous, one-hundred-acre retreat is within the Santa Monica Mountains National Recreational Area, about thirty-five miles northwest of where Miller's last Los Angeles shop still stands. If the ranch merits this distinction, should not the birthplace as well?

Robert M. Rubin

Bob Rubin shrugs off his role in this story by saying, "I think of myself as being more a fiduciary [a person who holds things in trust] than as an actor." Perhaps he's too close to the source of action. This young financier speaks hardly at all about himself and, although I have known him for some years now, I know little of his personal story. But I do perceive him as a front-center-of-stage actor, in the sense of his being the instigator of key events that generate whole chains of fascinating, constructive consequences. One such chain began with his acquisition in about 1988 of the ex-Harrah Miller 91 front-wheel-drive Packard Cable Special. When he went to Yale not so long ago, his major was history. He saw that car through the eyes of a professional in that discipline and, true or not, what has followed has had the air of having been inexorable. He confided the car's restoration to top expert Christopher F. Leydon, with brilliant results. Then came static displays at the annual Bagatelle Concours in Paris, as well as at Pebble Beach, California. Then came operational outings at Bridgehampton (Bob also acquired that shrine of historic-car racing along the way) and at Laguna Seca. Then came his donation of the car to the Smithsonian—which is to say to the American people—followed by the monograph, *Miller*, in which he wished that I should write about that American genius and his handiwork.

It was during the early growth of Bob's interest in things Miller that he sought out Buck Boudeman, and became acquainted with that unique guy and his wondrous collection. Those contacts took place around 1986. Rubin was smitten by the infinitely rare Four Wheel Drive Auto Company's four-wheel drive, which still was in quite decent running condition. He told Buck, "If you ever should decide to let that car go, I'd be very grateful if you'd let me know." Which is what happened, in about 1989. Bob told me, "Having communed with the 91, I was ready to plunk down another seven-figure sum for such an important and interesting machine. That archive of papers was a potent component of the car's attraction for me. I doubt that such complete documentation on the history of a single race car exists outside of the Mercedes-Benz factory. If there."

Millers are not the only steeds in Rubin's stable, nor the only irons in his fire. The four-wheel drive has been on a back burner while Chris Leydon has been giving his treatment to Bob's ex-Indianapolis 8CTF Maserati. That car's post-restoration debut was at Dave Uihlein's big Miller and other Championship car meet on the Milwaukee Mile oval on July 16, 1995. Following that, the FWD Special became the center of attention at Leydon Restorations, Ltd., the results of which, reported in the following chapter, bring the story of the 1932 Miller FWD Special up to press-time.

Chapter 20

Dean Butler: Back to Racing

The FWD Miller was restored at Leydon Restorations, Lahaska, Pennsylvania, from 1994 through mid-1996. When Leydon received the car, it was exactly as Bill Milliken had returned it to the FWD Museum in 1951—and in running condition—thanks to Buck Boudeman, who had been able to get it running with little effort after he acquired the car from the museum. The car was in remarkably original configuration, and still running with the 255 Miller-Offenhauser installed by Frank Brisco for the 1934 Indianapolis 500. Thus, the restoration was a reasonably straightforward job of totally disassembling the car and putting it back together in the best possible condition for a return to racing.

Illustration 20-1 The FWD driving the Colorado Grand 1996. (Bob Dunsmore Photography)

The car's first outing was not a race, but a remarkable test of driveability and reliability—the 1996 Colorado Grand. This 1000-mile, four-day run through the Colorado mountains in cooling September weather was a wonderful triumph for a wonderful race car and its restorers. Chris Leydon and his wife Rita piloted the Miller FWD the entire distance without incident. The car proved remarkably "streetable" and reliable—a real testimony to Miller, Goosen, Offenhauser—and Leydon. And, perhaps most importantly, this run proved that Miller's original intent, that of using the Miller FWD basic engineering concept as a launching platform for high-performance street cars, was perfectly feasible.

One of the more remarkable happenings on the Colorado Grand was that Chris Leydon, who had not personally rebuilt the gearbox/transfer case unit,* mistakenly presumed that the lever for engaging and disengaging the center differential was the one for changing the drive from four wheel drive to front wheel drive. Thus, with a mind toward minimizing tire and drive train wear, Leydon drove the entire Colorado Grand in what he thought was front-wheel drive. It was only after the entire 1000 miles had been driven that Leydon realized the car had been in four-wheel drive all along (with the center differential operating). I tell this story, at the modest risk of embarrassing one of the world's most talented race car restorers, to further demonstrate the tractability of Miller's four-wheel drive system. Chris Leydon has driven thousands of miles in all manner of higher-performance cars, and for Chris not to have sensed that all four wheels were driving the car on the Colorado Grand is more than adequate proof of the Miller design.

The car's second outing was at one of Bob Rubin's superb private race car party weekends at the now lamented historic Bridgehampton race track on the tip of Long Island. Bill Milliken came to see and drive the car (the first time since 1953) and Griff Borgeson came all the way from his home in France. Bill did not come fully prepared to actually drive the car. He arrived looking very much the dapper gentlemen that he is—in tan dress trousers, light yellow sweater vest, and sky blue sport coat. Griff arrived looking very much the casual sporting gentleman he always was—in jogging-type trousers and polo shirt. Bill had not necessarily expected to drive the car, but when offered the opportunity, quickly jumped into the car, blue sports jacket and all. Within only two laps, then 85-year-old Bill Milliken was driving the FWD Special at racing speeds—coming down the Bridgehampton straight at well over one hundred miles per hour. Then, after a few laps at speed, Bill came into the pits, and Griff was soon sitting beside him in the passenger's seat.

The sight of Bill Milliken, blue sports jacket flapping in the more than one-hundred-mile-per-hour air stream, with Griff Borgeson at his side, silver ponytail protruding from his racing helmet and accompanying Bill's jacket in the wind, is a scene forever emblazoned on my mind.

Others drove the car at Bridgehampton, including myself. My first impression was surprise at the massive torque and breadth of the torque curve. The three-speed gearbox was never designed for constant sports-racing-car-type shifting, and I thus found shifting silently rather difficult at first—but I also found that a lot of gear shifting was not really necessary. The

* One of the experts in Leydon's shop had actually done the work.

Illustration 20-2 Bill Milliken and Griff Borgeson with the FWD at Bridgehampton in 1996. (Frank Gudaitis)

engine pulled with massive torque from about 2500 rpm all the way to our self-imposed 5500 rpm red line. The car's handling on the bumpy Bridgehampton track was excellent—virtually as good as that of a T51 Bugatti or 8C Alfa of the same period. However, cornering was not in the tail-out style of the sports racers of the period. Rather, the car wanted to drift all four wheels without the tail hanging out.

As Bill Milliken had recommended with the track dry, the vast majority of our driving was with the center differential operating. We did, however, try the car with the center differential locked. The most noticeable difference was under braking—when the car hopped about alarmingly on the bumpy track. Bill Milliken, and his son, Doug, pointed out that this was caused by the resultant reduction of the front brake bias that would otherwise exist. With the center differential locked, the front and rear wheels were forced into braking harmony instead of the front bias.

One of those who drove the FWD Special at Bridgehampton was noted English vintage car racer Mark Gillies. Mark's first comment as he exited the car was, "I could beat Alfa Monzas with this." This would place the Miller's capabilities among the better sports racing cars of its period.

One of the Miller FWD's more interesting outings was a trip to the Goodwood Festival of Speed in Chichester, England, in June 1997—with then eighty-six-year-old Bill Milliken as driver. Goodwood is a one-car-at-a-time speed trial up the gently sloped, slightly more than one mile private drive of Goodwood House, the Earl of March's private residence. The

Miller FWD was in a class with about a dozen other Indianapolis racing cars, ranging from a 1923 Bugatti to a 1965, ex-Jim Clark Lotus 38, with the FWD being the second oldest Indy car.

For the first time in Goodwood Festival of Speed history, it rained—and rained—and Bill Milliken and the FWD Special were in their element. Even at 86 years of age, Bill took this event very seriously. Initially, Bill was expected to drive as a "demonstration," which means untimed on the hill. But Bill came to compete—not demonstrate. After a bit of cajoling and arm-twisting the officials agreed to time Bill's ascents on the hill—to the embarrassment of those in much newer Indy cars. When all was said and done, Bill achieved the second fastest time of all the Indy cars, having been beaten only by a 1950's Indy race car.

Illustration 20-3 Bill Milliken at the start line at Goodwood, June 1997. (Steve Havelock)

The FWD Special traveled back to the UK in 1998 for three events: the prestigious Louis Vuitton Concours at the Hurlingham Club in London, a major Vintage Sports Car Club race meeting at Donington, and a second showing at Goodwood.

At the Louis Vuitton Concours, the car was honored with the Rolls Royce Award for Engineering Excellence. At Donington, I drove the car myself—and qualified mid-field during a wet practice session, the four-wheel drive providing a nice advantage over a number of normally faster cars driven by more experienced drivers. But by race time the track had dried out. At the end of the first lap, on the back straight, just before the very tight Goddards right hand turn, the pack was traveling considerably faster than in the wet—including myself in the FWD. When it came time to brake, I discovered one of the car's shortcomings under road-racing conditions—and was forced to travel straight down the escape road, while the European racers slowed more quickly and disappeared around Goddards. By the time I had turned the FWD Special around, I was in last place. Within a few laps I was catching the back markers—when the race was red flagged due to an ERA having lost a wheel and parking on the edge of the track in an impact zone. What this event demonstrated to me is that Mark Gillies was correct when he said the car would be competitive with pre-war European road race cars.

The car's second Goodwood outing did not go well. Coming DOWN the hill after my second practice run, the engine stopped running. I tried to restart the engine while rolling down the hill at perhaps 30 miles per hour. It would not restart. Then I noticed the horrifying orange light of flames around my feet. The car had caught fire because a well-worn intake valve failed to seat. Trying to restart the engine while the car was rolling pressurized the intake manifold and pushed fuel out of the air intakes of the carburetors! Fortunately, the fire was extinguished before any harm was done. The lesson learned was that the old Miller-Offenhauser engine cannot be expected to be reliable for several thousand miles and requires more frequent internal inspection! The engine had not been apart since Leydon's rebuild.

The Miller-Offenhauser engine was then rebuilt in the UK—by a noted engine builder familiar with a broad range of vintage racing engines, including early post-war Jaguars. The engine builder was amazed to find that the valves and valve springs were dimensionally identical to early post-war Jaguar XK components. The valves and springs were replaced with stock Jaguar units. One wonders whether or not Jaguar looked at, and copied, these aspects of the 255 Offy valve train. The Offy certainly would have been regarded as one of the top designs to study.

Illustration 20-4 Bill and Doug Milliken at the Watkins Glenn 50th anniversary. (Michael Yust)

After its rebuild, the FWD Special was returned from England to the U.S.A. in late summer 1998—in time for Bill Milliken to demonstrate the car at the Watkins Glen 50th anniversary celebrations, where the car performed flawlessly for several tours of the old race course.

In July 1999, the car spent several uneventful hours circling the Milwaukee Mile during the Miller Reunion event—Bill Milliken once again one of the drivers, this time for a *Road & Track* magazine article.*

Plans are to continue using the FWD Special several times a year—for demonstrations and vintage racing events—continuing the car's lively history!

Dean Butler
July 1999

*"Salon," *Road & Track*, December 1999.

Epilogue

By Bill Milliken

In the course of his career, Harry Miller produced outstanding rear-, front-, and four-wheel-drive racing cars, in that order. Recognizing Miller's genius as a creative mechanical designer/artist, Griffith Borgeson has recorded the Miller saga in *The Golden Age of the American Racing Car* and in *Miller*, and now completes the trilogy with the present volume, *The Last Great Miller*.

The most successful example of a Miller four-wheel drive, based on its racing record, is still in existence, owned by E. Dean Butler and faithfully restored and maintained. The authenticity of this book is in no small part due to Borgeson's access to this machine through the sponsorship and cooperation of Butler, and the former owner, Robert Rubin.

The machine in question, as this book recounts, was originally commissioned by The Four Wheel Drive Auto Company (FWD) of Clintonville, Wisconsin to demonstrate the four-wheel-drive principle at high speed. Hence it competed at Indianapolis on a number of occasions, finally finishing in fourth place in 1936, driven by Mauri Rose. As also recorded in this book, the car was loaned to the present writer to compete in various hill climbs and post WWII road races.

In late October 1996, I received a phone call from Dean Butler inviting me to attend a private weekend session at Bridgehampton to again drive the FWD Miller. Rubin and Borgeson as well as Dean would be there. Inasmuch as I hadn't driven the Miller in nearly 50 years, I approached the weekend with, shall we say, some trepidation. Arriving at the track, I introduced myself to Dean and inquired, "What's this machine currently worth, I might bend it?" To which Dean responded, "Well, someone offered me a million-and-a-half, but don't worry, get in and drive it!" Such enthusiasm is infectious and set the tone of the weekend.

After a few laps I was comfortable again with the 3-speed, widely spaced crash box, double-clutching both down and up as had been my practice. We experimented with the effect on handling of locking up the center differential on dry, high-coefficient surfaces. Dean noted a "squirrelly" feeling under braking, which I also experienced, and which we have analyzed as effectively moving the brake distribution aft.

But apart from these technicalities, it was the hours of conversation with Griff Borgeson, Dean, and Bob that made the weekend memorable. In the years that I competed with the Miller, I recognized four-wheel drive as logical for traction-limited operation on a low-coefficient surface such as Pike's Peak, but I knew little of the fascinating details of the car's origin. As a master historian, Borgeson's practice was to seek out original sources, hundreds of them in fact, to find out why and how things actually happened. Likewise, Dean's interest extended far beyond the physical ownership of this classic car, to the circumstances of its history. I felt fortunate for having been exposed to these wider views.

In the months that followed, Griff and I were to carry on an extensive communication, written and verbal, as he produced those parts of the book which had to do with my particular experience with the FWD. I learned that Borgeson's interests extended in many directions including aircraft, and I came to appreciate the dedication that accompanied these interests. I had always been curious about Ader's L'Avion, and Griff immediately forwarded an authoritative technical description of it.

Early in 1997 I received another call from Dean, "I've been invited to enter the Miller in the Goodwood Festival of Speed, a hill climb event for classic race cars on the estate of the Earl of March—you gotta come over and drive it!" Having paused for perhaps a millisecond, I accepted!

The event was to take place in late June and Griff Borgeson was planning to attend. The draft of this, his Four-Wheel Drive book, was approaching completion and he was planning to cover the Miller's participation at Goodwood in the final section.

A driver at Goodwood has two alternatives—he can treat it as a demonstration (exhibit) by making untimed runs at comfortable speeds or he can drive it as a competitive speed hill climb event. Considering the thrilling history of this wonderful old race car, loafing up the hill seemed as incongruous as expecting a horse like Man of War to walk in front of a crowd. So we chose the latter, and in the months that followed engaged in the following preparations:

Using a map of Goodwood we divided the course into straights and turns, tabulating lengths and radii and also elevation changes. Next, we summarized the known and estimated data on the Miller's weight and CG (center of gravity), engine torque and power, the gearing, and the suspension characteristics that affect the ride and roll rates. Armed with these data, an assumed road surface friction, or cornering g's, and a redlined engine rpm (or piston speed), we entered our Lap Time Simulation computer program. This work was largely accomplished by Doug Milliken. Five sets of runs using different operating assumptions were made, but for learning purposes we used the case of 0.8g cornering acceleration and 0.6g braking with a redline of 5800 engine rpm—conservative values for a dry surface. The results are presented in Figure 1.

Because of the widely spaced gearing and the nature of the course, it appeared that most of the run could be made in second gear, with an estimated time of a little over a minute. For driver preparation we installed a "g·Analyst" recorder in our Chrysler New Yorker, and in a month of building up to a respectable cornering acceleration wore out the front pair of Goodyear Eagles!

When we were about to leave for England, we received word that Griff was confined by illness and would be unable to attend. This sad turn of events cast a shadow over a nearly perfect weekend. Had he been there the story of Goodwood would have been told with all its magical atmosphere and pageantry. The following notes merely cover some of the more factual aspects.

Traveling with Tony Best, an old friend and business associate, we met up with Dean at Stakis Arundel Motel, our headquarters near Goodwood on Thursday before practice. It was a weekend of rain, ideal for showing off the attributes of four-wheel drive by merely locking up the center differential, which goes a long way in inhibiting wheel spin. Between

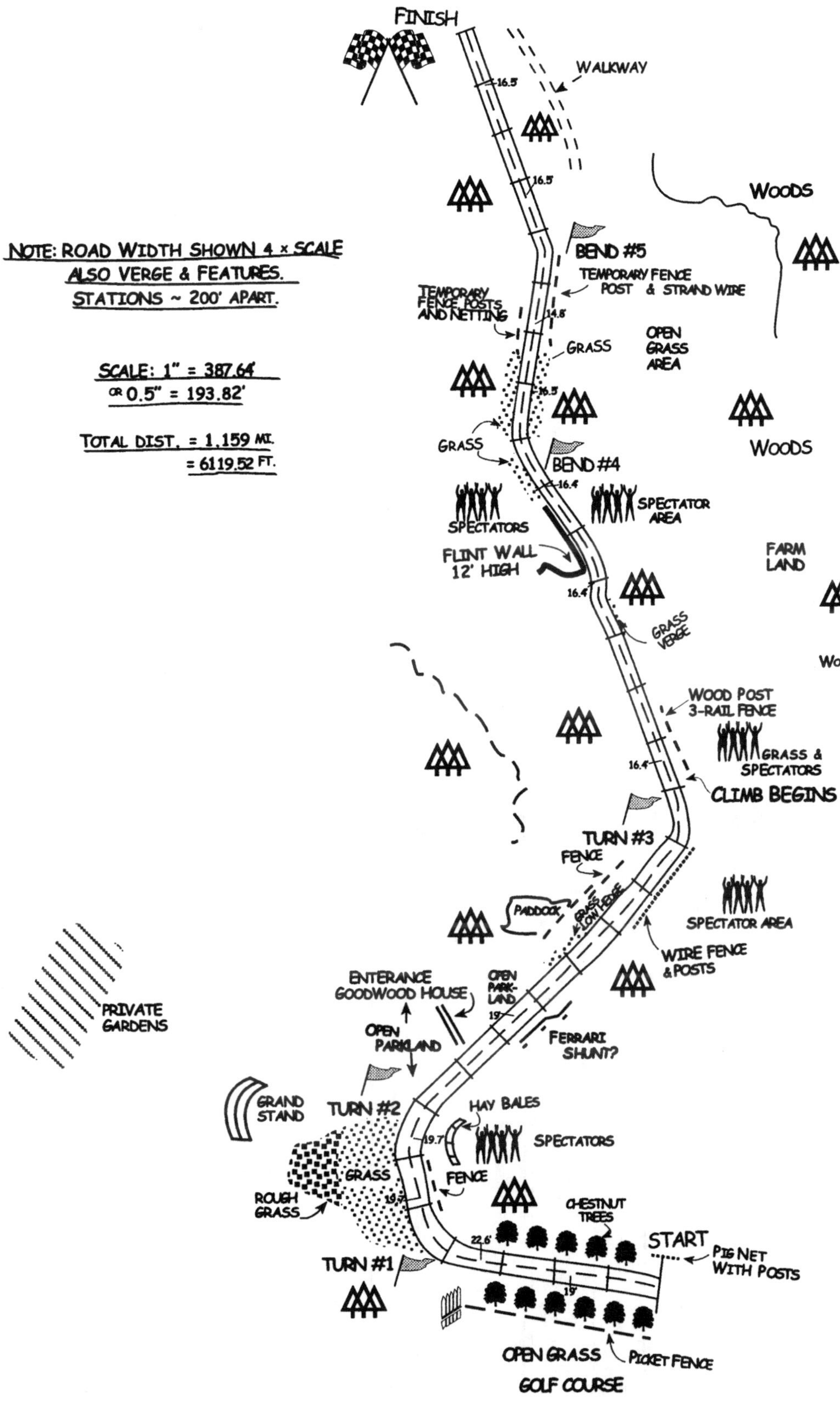

Figure 1 The hill climb course with its three turns and two bends. It is broken into 200-foot segments for analysis.

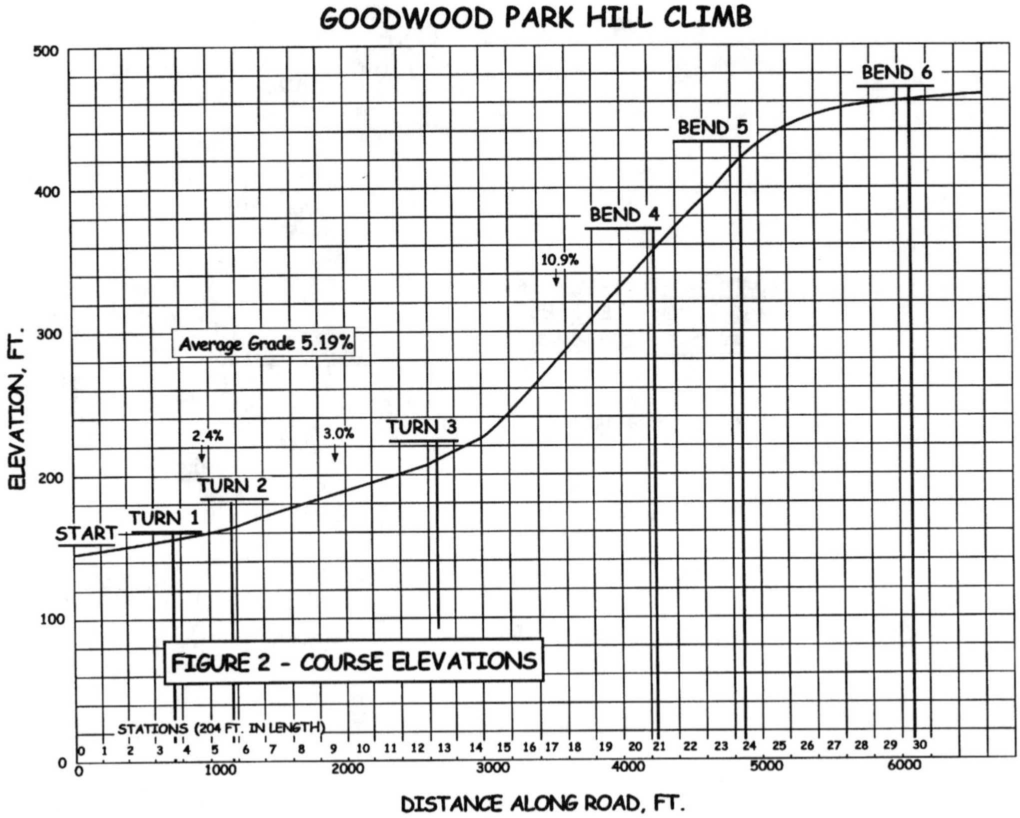

Figure 2 The elevations plotted against distance along the course. From an initial grade of about 3% the road pitches steeply up to nearly 11%.

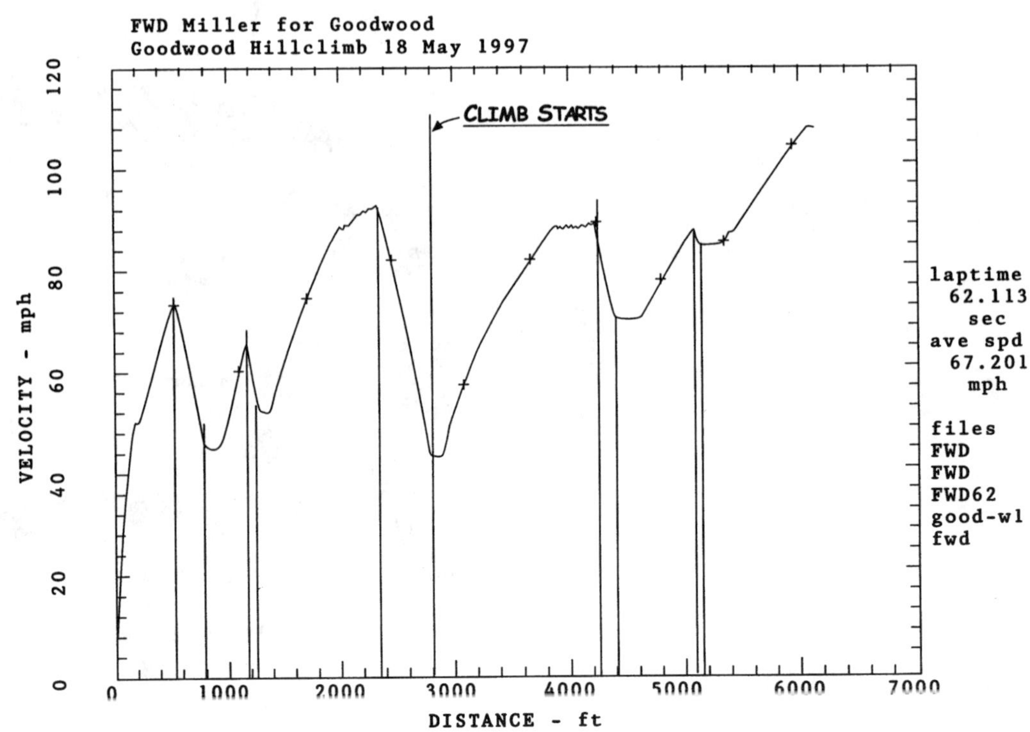

Figure 3 The calculated speed trace, again plotted against distance along the course.

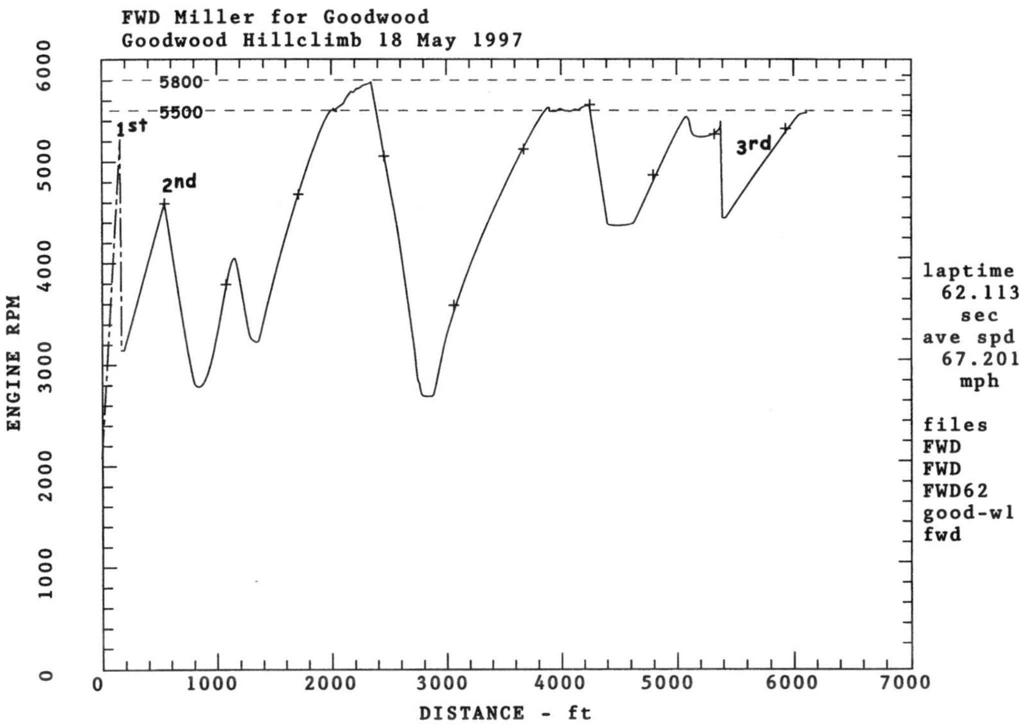

Figure 4 The calculated engine RPM and the gears used.

practice and the event itself we made eight runs up the hill. On the first run, I was surprised to find that Turn No. 3 is much tighter in actuality than we assumed for the calculations, but otherwise the calculations were a useful guide. As my friend, John Fitch had predicted, the approach to Bend No. 4 with its 12-foot-high flintstone wall on the left is deceptive, if not scary. Charging up the pitch, the wall appears to be almost straight ahead and one must avoid the tendency to momentarily lift off.

With the rain another problem was encountered. The electrics misbehaved so that the on-board starting system failed to work. It became necessary to push or tow the car. In my experience, the performance of the crew members is paramount to success; they not only prepare the car, but inspire the driver to do his best. In this case it was Keith Chapman, Martin Wolford, Mark Turner, Paul Myatt, and Tony Best who made it possible to do the runs on schedule.

What is it like to drive the Miller? By design the four-wheel-drive Miller is a neutral steerer. Thus, the CG is at the neutral steer point very close to mid-wheel base; the suspensions (deDion) are identical front and rear and the roll axis is horizontal; front and rear tires and pressures are normally the same, and under road load conditions front and rear horizontal forces are equal. The result is that the machine corners on the Ackermann and feels as if it is on a track. Under heavy traction or braking there is a substantial longitudinal weight transfer due to the high CG and like all cars of that era, it possesses trailing throttle oversteer and wants to spin when approaching and braking-in-a-turn. If one is cornering under large traction and backs off on the throttle, the car will "duck" into the turn and perhaps spin out if corrective steering is not applied. Cornering with center differential "operating" on wet or low coefficient surfaces can result in push or spin depending upon the direction of the longitudinal weight transfer, but locking up the center diff goes a long way in ameliorating this behavior. Thanks to its extremely stiff suspension (220 cycles/minute or twice as stiff as most classic race cars), the FWD Miller is as alive as a Kart and an exciting handful on rough surfaces, as I had found out on Mt. Equinox. The steering is good, but with low returnability.

Thus, like all race cars, the Miller has its own set of handling characteristics, which when learned make for an effective and intriguing machine, particularly in a hill-climb environment. My most thrilling moment was on the last run—having negotiated the wall with hardly a lift, we went over the finish line flat out in 2nd, the engine screaming at nearly 6000 rpm. I was hoping for a good trap speed, but the timing equipment failed! All and all we did well in the Indianapolis car class.

On a personal note, it was an occasion to reexperience some of my fondest memories, which go back to Pike's Peak, Watkins Glen, and Mt. Equinox of nearly fifty years ago. It was an opportunity afforded few people, and I'd like to come back at the end of the next fifty!

After the event, on July 4, I faxed Griff describing the weekend at Goodwood. With immense sadness I learned that Griff had passed away a few days earlier. Thus, this book, as the last of his great series of motor racing histories, stands as a memorial to Griff. His character, knowledge, and writings have enriched the lives of so many people worldwide.

Bill Milliken
June 24, 1998

Index

About the Author

Griffith ("Griff") Borgeson was born 21 December 1918 and died in the South of France on 29 June 1997, shortly after completing the first draft of this book. Throughout his life, he had written over 15 books and hundreds of magazine articles on automotive history. He was long recognized as one of the world's preeminent automotive historians.

Once Editor-in-Chief of *Motor Trend Magazine*, Borgeson also wrote for *Sports Car Illustrated Magazine* (now *Car & Driver*) and was the European editor for *Automobile Quarterly*. He also worked as a consultant for the Lincoln Division of the Ford Motor Company. The following list is a sample of his many writing awards: In 1966 the first edition of *The Golden Age of the American Racing Car* won the Thomas McKean Award for "book of the year" from the Antique Automobile Club of America; The Society of Automotive Historians Cugnot Award for "the outstanding magazine article of 1980" for "Beginnings of the VW Beetle" published in *Automotive Quarterly*; The Society of Automotive Historians Benz Award for "the outstanding magazine article of 1984" for "Edmund Rumpler: Icarus Bound" published in *Automotive Quarterly*; and The ITALA Prize from the Associazone Italiana Per La Storia Dell'Automobile (AISA) in 1985 for his collected works regarding the history of the Italian automobile.

One of his great strengths as an author was that he established personal relationships with many pioneer builders, designers, and race car drivers. His intimate knowledge of the development of automotive technology could have come only from his relationships with the greats of automotive history. He shared information with his readers that could not be gleaned simply from history books. This among other traits set him apart from other automotive writers.

Some of his other books include: *The Classic Twin-Cam Engine* (Dalton Watson, 1981); *Bugatti by Borgeson*, (Osprey Publishing, 1981); and, in conjunction with the Smithsonian Institution, *Miller* (Motorbooks International, 1993).